Battery Management System for Future
Electric Vehicles

Battery Management System for Future Electric Vehicles

Editors

Dirk Söffker
Bedatri Moulik

MDPI • Basel • Beijing • Wuhan • Barcelona • Belgrade • Manchester • Tokyo • Cluj • Tianjin

Editors
Dirk Söffker
University of Duisburg-Essen
Germany

Bedatri Moulik
Amity University
India

Editorial Office
MDPI
St. Alban-Anlage 66
4052 Basel, Switzerland

This is a reprint of articles from the Special Issue published online in the open access journal *Applied Sciences* (ISSN 2076-3417) (available at: https://www.mdpi.com/journal/applsci/special_issues/battery_management_vehicle).

For citation purposes, cite each article independently as indicated on the article page online and as indicated below:

LastName, A.A.; LastName, B.B.; LastName, C.C. Article Title. *Journal Name* **Year**, *Article Number*, Page Range.

ISBN 978-3-03943-350-6 (Hbk)
ISBN 978-3-03943-351-3 (PDF)

Contents

About the Editors

Dirk Söffker Professor—Chair of Dynamics and Control (SRS), University of Duisburg-Essen, Duisburg campus, Germany. Dirk Soeffker received a Dr.-Ing. degree in mechanical engineering and a Habilitation degree in automatic control/safety engineering from University of Wuppertal, Wuppertal, Germany, in 1995 and 2001, respectively. Since 2001, he has led the Chair of Dynamics and Control with the Engineering Faculty, University of Duisburg-Essen, Germany. His current research interests include elastic mechanical structures, modern methods of control theory, human interaction with safety systems, safety and reliability control engineering of technical systems, and cognitive technical systems.

Bedatri Moulik Assistant Professor—Department of Electrical and Electronics Engineering, ASET, Amity University, Noida, India. Bedatri Moulik received her Bachelors degree in Electrical Engineering and Masters degree in Control and Instrumentation Engineering from West Bengal University of Technology, Kolkata, India. She received her Dr.-Ing. degree from the Chair of Dynamics and Control, University of Duisburg-Essen, Germany. She is now an Assistant Professor at Amity University, Noida, India in the department of Electrical and Electronics Engineering. Her current research interests include battery parameter estimation methods, battery and power management in electric/hybrid vehicles, drive pattern recognition and intelligent control in the context of electric/hybrid vehicles.

Editorial

Battery Management System for Future Electric Vehicles

Bedatri Moulik [1],* and Dirk Söffker [2]

[1] Department of Electrical and Electronics Engineering, Amity University, Sector 125, Noida 201313, India
[2] Chair of Dynamics and Control, University of Duisburg-Essen, Lotharstrasse 1-21,
    47057 Duisburg, Germany; soeffker@uni-due.de
* Correspondence: bmoulik@amity.edu

Received: 7 July 2020; Accepted: 21 July 2020; Published: 24 July 2020

Abstract: The future of electric vehicles relies nearly entirely on the design, monitoring, and control of the vehicle battery and its associated systems. Along with an initial optimal design of the cell/pack-level structure, the runtime performance of the battery needs to be continuously monitored and optimized for a safe and reliable operation and prolonged life. Improved charging techniques need to be developed to protect and preserve the battery. The scope of this Special Issue is to address all the above issues by promoting innovative design concepts, modeling and state estimation techniques, charging/discharging management, and hybridization with other storage components.

Keywords: battery electric vehicles; battery management; hybrid energy storage

---

1. Introduction

Recent advancements in battery technology have pushed the sales of electric (EVs) and hybrid electric vehicles (HEVs) further. The improvements in the EV/HEV range, energy/charging efficiency, safety, reliability, and lifetime are entirely dependent on the design and chemistry of the battery pack and its associated systems. Most of the safety concerns regarding the battery's unexpected temperature rise and predictions of the internal reactions leading to fluctuations in internal temperature also need to be addressed.

2. Battery Management System for Future Electric Vehicles

The aim of the Special Issue "Battery Management System for Future Electric Vehicles" is to investigate advanced battery management technologies for the estimation, monitoring, and control of battery states, associated modeling techniques, thermal and charging/discharging management for optimized life, performance, and range. Optimal sizing, the hybridization of storage systems, and innovative battery test-benches were also encouraged. There are a total of seven accepted and published papers, which are summarized as follows:

The first paper, authored by Hakeem and Solyali [1], presents a battery thermal management system (BTMS) with improved performance in terms of battery cooling. An improved pack structure is proposed, which is experimentally investigated with different air flow rates and current rates of charge–discharge profiles. Finally, based on the obtained data, an artificial neural network is trained to obtain the thermal model of the battery pack.

The second paper, authored by Tseng and Yang [2], presents a torque and battery distribution strategy (TBD) that takes into account the torque–speed characteristics, as well as the battery state of charge to obtain optimized range and efficiency. Based on the State of Charge (SoC) gaps and ratios setween the front and the rear battery packs, three torque distribution modes are then proposed. First simulation, then hardware-in-the-loop experimentation, followed by actual road tests, are performed to validate the effectiveness of the TBD in the extension of the electric vehicle range.

The third paper, authored by Kuo [3], presents a battery model based on a modified Thévenin circuit, Butler–Volmer kinetics, Arrhenius equation, Peukert's law, and a back propagation neural network (BPNN). The model can estimate the coulombic efficiency and the remaining capacity of the battery, as analyzed experimentally under various environmental conditions. Based on experimental results and curve fitting techniques, a comprehensive model is developed. A correction factor is introduced and the prediction of remaining capacity is done using a BPNN.

The fourth paper, authored by Cao [4], presents a wireless distributed and enabled battery energy storage system (WEDES) for electric vehicles (EVs) derived using a small signal modeling technique. The WEDES controller is designed to address SoC balancing, bus voltage regulation, and battery module current/voltage regulation at the same time. Finally, simulation and hardware experiments are carried out to evaluate and validate the accuracy and effectiveness of the derived model and controller.

The fifth paper, authored by Guo, et al. [5], presents an online SoC estimation method by using an equivalent circuit model, followed by model parameter identification. An optimization method is proposed to improve the accuracy of the SoC estimation. Then, an online estimation based on the adaptive unscented Kalman filter (AUKF), and with optimized model parameters, is performed. The estimation accuracy of the AUKF with the UKF is compared. The convergence of the initial error of the AUKF before and after parameter optimization is also compared.

The sixth paper, authored by Chen, Chen, and Duan [6], presents an optimization method to cooperatively optimize the economic dispatching and capacity allocation of both renewable energy sources (RESs) and electric vehicles (EVs). Both the installation capacity of RESs and the number of EV charging/discharging infrastructures (EVCDIs) are optimized. This optimization method is based on the EVs' across-time-and-space energy transmission. The main optimization objective is to improve the economics of the system allocation and decrease the cost of the microgrid operator. A two-loop optimization is considered using an improved particle swarm optimization (IPSO). The inner loop comprises the optimization of system dispatching, while in the outer loop, the allocation of EVs and RESs is optimized.

The seventh paper, authored by Hou, et al. [7], presents a variational Bayesian approximation-based adaptive dual extended Kalman filter (VB-ADEKF) to improve the accuracy of SoC estimation. First, the variational Bayesian results are used along with the extended Kalman filter to jointly estimate the states. Next, both variational Bayesian and variational Bayesian approximation-based adaptive dual extended Kalman filters are alternatively used. Additionally, measurement noise variances are considered to compensate for uncertainties in measurement. With the help of experiments, the proposed VB-ADEKF algorithm is compared with the traditional DEKF algorithm in terms of SoC estimation accuracy, convergence rate, and robustness.

Thus, summarizing all the seven papers brings us to the conclusion that this Special Issue has been successful in bringing together novel contributions considering multiple aspects of energy storage management and optimized charging/discharging schedules.

3. Future Battery Management Systems

Although this Special Issue is finished, immense work still remains in the field of innovative battery state estimation algorithms and optimization approaches to improve their accuracy and reliability in terms of online and real-time application in EVs.

Author Contributions: Both authors contributed equal parts to the paper, whereby the corresponding author was mainly responsible for initial writing, the second author for structuring and organizing, reviewing, and summarizing the entire contribution. All authors have read and agreed to the published version of the manuscript.

Funding: This research received no external funding.

Acknowledgments: This Special Issue would not have been a success without the constant efforts of the authors—including those whose papers could not be selected. We can only hope that this Special Issue is able to provide them with a platform to learn and improve their existing works based on the valuable comments of the reviewers and editors. We would also like to thank the reviewers for dedicating their time and energy to the submitted manuscripts and motivating the authors to improve their work. Last, but not least, we would like to extend our heartiest congratulations to the entire editorial team of *Applied Sciences* for their sincere efforts and constant dedication in making this Special Issue a success.

Conflicts of Interest: The authors declare no conflict of interest.

References

1. Hakeem, A.A.; Solyali, D. Empirical Thermal Performance Investigation of a Compact Lithium Ion Battery Module under Forced Convection Cooling. *Appl. Sci.* **2020**, *10*, 3732. [CrossRef]
2. Tseng, Y.H.; Yang, Y.P. Torque and Battery Distribution Strategy for Saving Energy of an Electric Vehicle with Three Traction Motors. *Appl. Sci.* **2020**, *10*, 2653. [CrossRef]
3. Kuo, T.J. Development of a Comprehensive Model for the Coulombic Efficiency and Capacity Fade of LiFePO4 Batteries under Different Aging Conditions. *Appl. Sci.* **2019**, *9*, 4572. [CrossRef]
4. Cao, Y. Small-Signal Modeling and Analysis for a Wirelessly Distributed and Enabled Battery Energy Storage System of Electric Vehicles. *Appl. Sci.* **2019**, *9*, 4249. [CrossRef]
5. Guo, X.; Xu, X.; Geng, J.; Hua, X.; Gao, Y.; Liu, Z. SOC Estimation with an Adaptive Unscented Kalman Filter Based on Model Parameter Optimization. *Appl. Sci.* **2019**, *9*, 4177. [CrossRef]
6. Chen, J.; Chen, C.; Duan, S. Cooperative optimization of electric vehicles and renewable energy resources in a regional multi-microgrid system. *Appl. Sci.* **2019**, *9*, 2267. [CrossRef]
7. Hou, J.; Yang, Y.; He, H.; Gao, T. Adaptive dual extended Kalman filter based on variational bayesian approximation for joint estimation of lithium-ion battery state of charge and model parameters. *Appl. Sci.* **2019**, *9*, 1726. [CrossRef]

Article

Empirical Thermal Performance Investigation of a Compact Lithium Ion Battery Module under Forced Convection Cooling

Akinlabi A. A. Hakeem [1] and Davut Solyali [2,*]

[1] Department of Engineering, Faculty of Mechanical Engineering–Eastern Mediterranean University, Famagusta, North Cyprus 99628, Turkey; abdulhakeemridwan@gmail.com or 17500481@students.emu.edu.tr
[2] Director, Electric Vehicle Development Center (EVDC), Eastern Mediterranean University, Famagusta, North Cyprus 99628, Turkey
* Correspondence: davut.solyali@emu.edu.tr

Received: 10 April 2020; Accepted: 16 May 2020; Published: 28 May 2020

Abstract: Lithium ion batteries (LiBs) are considered one of the most suitable power options for electric vehicle (EV) drivetrains, known for having low self-discharging properties which hence provide a long life-cycle operation. To obtain maximum power output from LiBs, it is necessary to critically monitor operating conditions which affect their performance and life span. This paper investigates the thermal performance of a battery thermal management system (BTMS) for a battery pack housing 100 NCR18650 lithium ion cells. Maximum cell temperature (Tmax) and maximum temperature difference (ΔTmax) between cells were the performance criteria for the battery pack. The battery pack is investigated for three levels of air flow rate combined with two current rate using a full factorial Design of Experiment (DoE) method. A worst case scenario of cell Tmax averaged at 36.1 °C was recorded during a 0.75 C charge experiment and 37.5 °C during a 0.75 C discharge under a 1.4 m/s flow rate. While a 54.28% reduction in ΔTmax between the cells was achieved by increasing the air flow rate in the 0.75 C charge experiment from 1.4 m/s to 3.4 m/s. Conclusively, increasing BTMS performance with increasing air flow rate was a common trend observed in the experimental data after analyzing various experiment results.

Keywords: air-cooled BTMS; electric vehicle; compact lithium ion battery module; ANN

---

1. Introduction

There is a growing global concern of the causes and effects of climate change which has led to stricter environmental regulations on carbon-based machines [1,2] coupled with huge advancements in portable battery technology—specifically, lithium ion electric vehicles (EVs) and hybrid electric vehicles, which are starting to disrupt the automobile industry markets by presenting themselves as the vehicle choice of the future [3,4]. Some major hindrances to electric vehicle mass adaptation are the range anxiety of EVs, the lack of super-fast charging and the lack of performance driving, etc. [1,5]. The performance driving and fast charging problems of EVs are due to the limitation of the lithium ion batteries in performing outside tight operating temperature ranges [6]. The range anxiety problem of electric vehicles is also attributed to the gravimetric density of lithium ion batteries (LiBs). When compared to traditional gasoline-powered vehicles, the average energy-to-weight ratio of lithium ion batteries is 0.3 MJ/kg and it is over 30 MJ/kg for gasoline-powered vehicles [7].

While the current gravimetric property limitation of LiBs may be a design constraint on EV performance, EV manufacturers have the freedom to design robust battery thermal management systems (BTMS) for EV battery packs (BP) to efficiently limit the amount of heat generated by the LiBs during their operating cycles (charge/discharge). One technique; a sub-classification (see Figure 1) of air-cooled BTMS employed by various researchers used in improving the cooling performance of a BTMS, is reviewed and investigated in this paper.

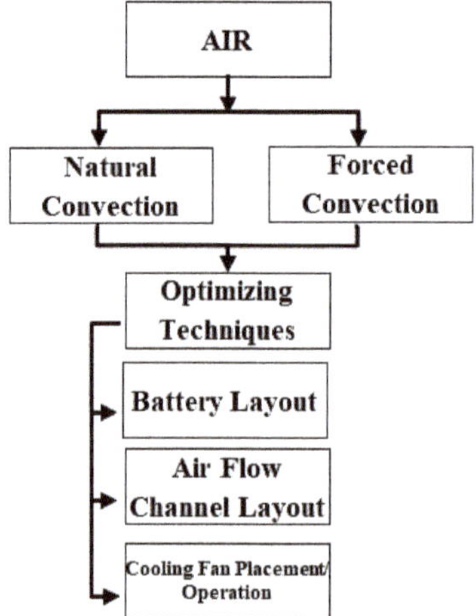

Figure 1. Classification of air-cooled battery thermal management systems (BTMS) and optimization parameters adapted from [1,4,8].

1.1. Literature Review

In recent years there have been many studies performed with the aim of improving the cooling performance of air-cooled BTMS by employing optimizing techniques as illustrated in Figure 1. The research studies conducted regarding the performance improvement of BTMS which deal primarily with design variables pertaining to manners of arranging cells inside a BP and the placement of cooling-air intake and exhaust vents to obtain the best performance are discussed in the following paragraphs.

Chen et al. (2017) performed a configuration optimization on prismatic lithium ion cells for a parallel air-cooled system. In this model, the BTMS is optimized through arranging the spacing among the battery cells to obtain the best cooling performance. The optimization strategy is applied several times on a developed flow resistance model and a heat transfer model until the appropriate cell spacing is obtained. Their results exhibited a 42% reduction in the maximum cell temperature over the design variable optimization iterations on the developed model [9].

By comparing an aligned versus staggered cylindrical cell arrangement for a BTMS (see Figure 2A,B), N. Yang et al. (2015) in [10] investigated the effects of transverse and longitudinal spacing between cylindrical cells in a BP with a forced-air cooling system. N. Yang et al. (2015) developed a numerical and thermal model for this BP, which was used to simulate the effects of various design variables on their BP model. The model with the best performance results was validated by physical experiment. N. Yang et al. (2018) reported that under a specific cooling-air flow rate, the maximum cell temperature rise in a BP is proportional to the longitudinal interval for staggered arrays, whereas the inverse holds for aligned cell arrays. Finally, they obtained a better performing BTMS model, by optimizing the longitudinal and transverse space between the cells, coupled with optimizing an air inlet duct width for a BP with aligned arrangements [10].

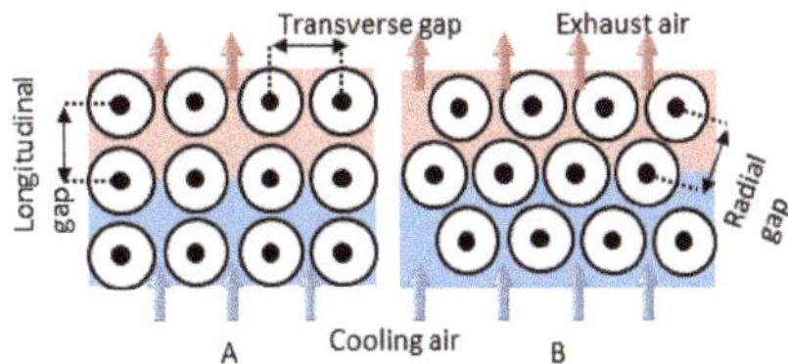

Figure 2. Schematic of aligned (**A**) and staggered (**B**) cell layout optimization adapted from [4,10].

Lu (2018) provided a parametric study of forced-air cooling for lithium ion batteries with staggered arrangements. They designed a three-dimensional simulation model (Gambit 2.4.6 CFD) of the BP that investigated the effects of cooling channel size and air supply strategy for the model. The CFD model was solved using the semi-implicit method for pressure-linked equations (SIMPLE) algorithm. They deduced that a cooling channel size of 1 mm was appropriate for BPs with Panasonic 18,650 cells. Upon further investigation, they reported the best cooling performance was achieved when placing the cooling-air flow inlet and outlet on the top of the BP. Finally, they reported that the efficiency factor of a BTMS deceases with the number of cells in the horizontal direction; hence, they recommended a maximum of 10 cylindrical cells along the air flow direction for a BP [11].

In a more recent study, Chen et al. (2020) in their paper numerically studied five (5) BTMS battery pack configurations and verified simulation results by conducting physical experiments. They developed a simple method to achieve symmetrical air flow inside each of the five battery packs by repositioning inlet and outlet vents on each original battery pack design (I, II, III, IV and V) to get newly optimized BPs; (I1, II1, III1, IV1 and V1) as depicted in Figure 3 below. Further parametric optimization of cell spacing revealed that uneven cell spacing in the improved battery packs resulted in better BTMS cooling performance just as BPs with a symmetrical air flow path did over their original counterparts [12]

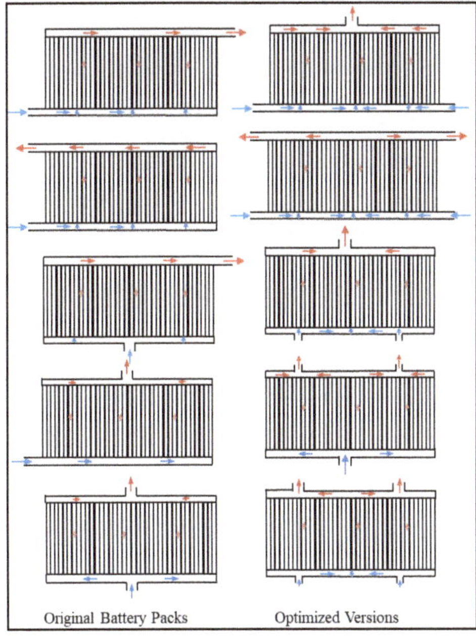

Figure 3. Asymmetrical vs. symmetrical BTMS battery packs adapted from [4,12].

1.2. Current Study

This paper investigates the performance of a battery thermal management system (BTMS) for a proposed battery pack model with "H" symmetrical air vents housing 100 NCR18650 lithium ion cells. The battery module is built in a 10P10S configuration with aligned cells. Maximum cell temperature (Tmax) and maximum temperature difference (ΔTmax) between cells were the performance criteria for the BTMS. The battery pack is investigated for three levels (1.4 m/s, 2.4 m/s and 3.4 m/s) of air flow rate combined using a full factorial experiment design method with two current rates (0.5 C and 0.75 C) under charging and discharging power cycles.

2. Investigated Battery Module

The cells in the battery module (Figure 4 above) are connected in a 10S10P configuration to provide a minimum rated power of 1.024 KWh minimum and 1.344 KWh maximum. Nominal data specification value of a single NCR18650 cell used in the battery module being tested are presented in Table 1 below.

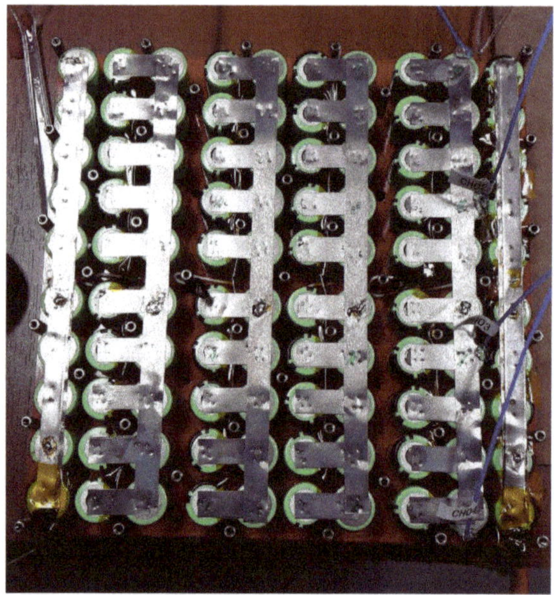

Figure 4. Battery module investigated.

Table 1. NCR18650B (green) Specification [13].

Specifications	Value
Nominal Voltage	3.6 V
Cutoff Voltage	4.2 V
Minimum Rated Capacity	3.2 Ah

An experimental setup up was designed to test the battery module for real-life scenarios while varying optimizing parameters—cooling-air velocity and current flow rate of the battery pack.

The experimental setup in Figure 5 consists of a battery pack with four switch-mode power supply (SMPS) cooling fans, a 3D-printed part connecting a flexible vent pipe to connect the fans to the atmosphere, a CPX400D (Aim TTi, Huntingdon, United Kingdom) power supply, a 16-channel temperature data acquisition device (Applent, China), a 4000 W EA-EL 9200 electronic load (Electro Automatik, San Diego, CA, USA) and a 200 V EA-PSI 9200 (Electro Automatic, San Diego, CA, USA) power supply.

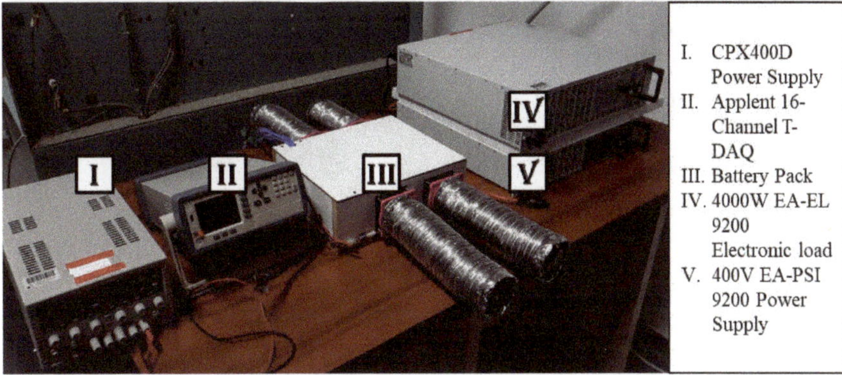

Figure 5. Experimental setup.

2.1. Experimental Setup

2.1.1. Battery Pack

The battery pack in the experimental setup was designed and built from medium fiberboard (MDF) of 18 cm thickness with four vent holes of 0.07 m diameter (Figure 6a). The pack had dimensions ($30 \times 25 \times 10$ cm) and was built slightly larger than the exact volume of the battery module ($24 \times 24 \times 10$ cm). This extra volume in the MDF box was designed to create a partition that would accommodate the excess length of the thermocouple sensor wires attached to the cells in the battery module (see Figure 6b). Glass fiber, (a nonconductive and nonflammable material) was placed between the battery module top cover of the MDF board as a protective measure to prevent possible electrical and fire hazards present during the experiment and most importantly prevent a low-resistance path for air flowing into the battery pack (Figure 6c).

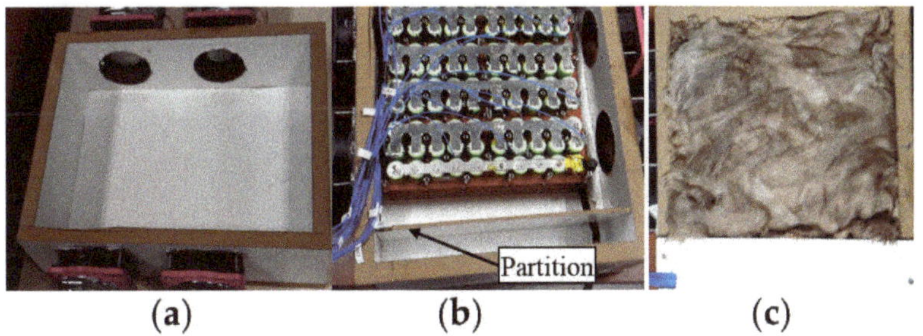

Figure 6. (a) Built battery pack from fiberboard, (b) Placement of thermocouples to the cells. (c) Protective wool placed on top of the battery cells.

2.1.2. Temperature Data Acquisitions Device

In order to monitor and record the temperature of cells in the battery module during testing, a 16-channel Applent temperature data acquisition device in Figure 7 was employed.

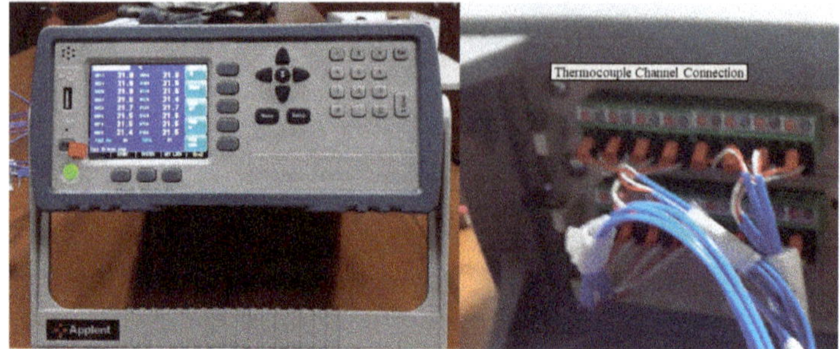

Figure 7. Temperature data acquisition device (T-DAQ).

The T-DAQ used in the experimental setup was powered by an ARM microprocessor capable of measuring temperatures from a variety of thermocouple types (T, J, K, E, etc.) at three sampling rates (fast, medium and slow) and had a resolution of 0.1 °C. It allowed for multiple recording of temperature values from 16 channels simultaneously which it stored onto a USB stick or directly onto a computer via a USB-serial connection.

K-Type thermocouples sensors with lower and upper limits of 0 °C and 200 °C temperature ranges well within the limits of the expected temperature rise of the cells in the battery module to be tested were employed during the experiment to measure temperature profiles of selected cells.

The positions of cells to be monitored was systematically selected based on the assumption that the temperature of each selected cell would represent the local temperature of other cells in its surrounding. Factors such as the number of channels the T-DAQ is limited to impacted the decision to monitor only 16 cells out of 100 cells in the battery module, as well as the preexisting compact nature of the battery module (see Figure 8 below).

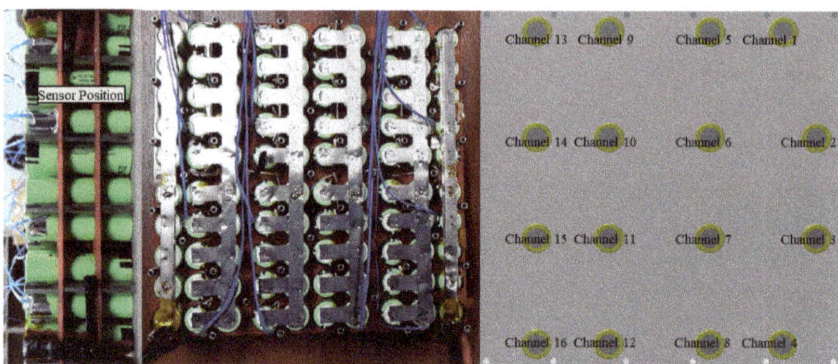

Figure 8. Thermocouple position selection and attachment.

The following measurements were taken to ensure proper contact between the sensor and the cells:

- In order to ensure proper cell-sensor connection and improve conduction between the two entities, a thermal paste compound was applied around the sensor and the battery body in contact after which strips of strong adhesive tape were used to secure the sensor to the body of the cell (see Figure 9).
- Silicone glue was applied to hold the protruding sensor wire at the attachment point on top of the cell terminal to provide for extra attachment strength (see Figure 9).

- Finally, the majority of the sensor wire length was kept folded in a separate partitioned section inside the MDF box to prevent accidental tension that might threaten or sever the connection between the sensor-cell attachment.

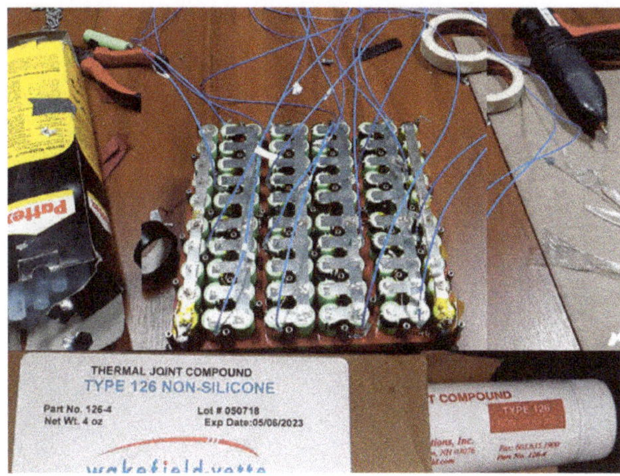

Figure 9. Sensor attachment measures.

2.1.3. CPX400D Power Supply

To achieve cooling on the battery module in the battery pack, four (4) SMPS fans attached to the vents of the MDF box were connected in parallel and routed via two connecting cables on the MDF box to be powered simultaneously using the CPX400D power supply Figure 10 below. This method of connection ensured all the cooling fans operated at the same speed at any preset voltage.

Operating the SMPS fans in constant current mode, controlling the voltage input to the fans via the power supply allowed for the adjustment of power delivered to the fans hence controlling air flow into the battery pack.

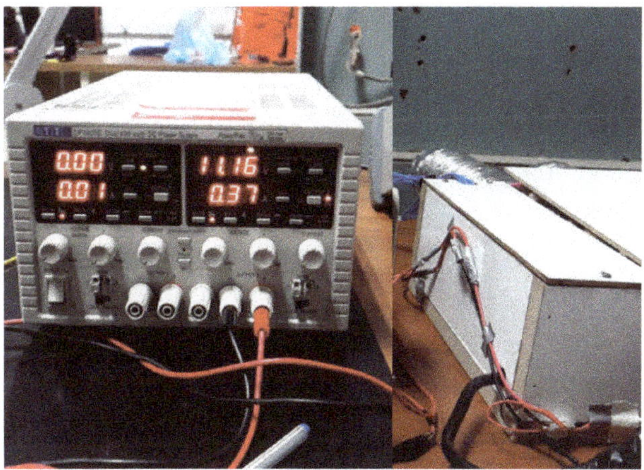

Figure 10. Controlling fan speed with the CPX400D power supply.

Figure 11 shows the base air-flow configuration of the battery pack investigated in this thesis.

Figure 11. Battery vent configuration, inlet Fans I and II, outlet fans III and IV.

2.1.4. Electronic Power Supply/Load

The electronic load and power supply used in this study were the Electro-Automatik (EA) heavy duty laboratory Direct Current (DC) load and power supply. The versions used in conducting the experiments—the EA-EL 9200 electronic load capable of an output of 4000 W and the EA-PSI 9200—operated at efficiency of up to 95.5% (see Figure 12).

To test the battery pack module for charge and discharge cycles during the experiments, 50 Amps-rated cables were used in connection between the battery pack, the electronic load and power supply.

Figure 12. EA electronic power supply and load.

As testing for two current ratings on the battery pack module (0.5 C and 0.75 C) were to be investigated, the load and power supply were programmed to a constant current rating of 0.5 C, and the maximum and minimum voltages set to fit the specification of the battery pack maximum and minimum cut off voltages (see Table 2). The Graphic User Interface (GUI) of the load/supply provided real-time voltage and current readings of the battery pack for monitoring purposes during the charge/discharge cycles.

Table 2. Battery Module Specification.

Ten (10) Cells in Series	Maximum Cutoff Voltage (V): **4.2 × 10 = 42 Volts**	Minimum Cutoff Voltage (V): **3.2 × 10 = 32 Volts**
Ten (10) Cells in Parallel	Nominal battery module capacity (Amp): 3.2 × 10 = 32 Amps	
Tested Capacity	0.75 C	24 Amps
	0.5 C	16 Amps

2.2. Designing of Experiments

This study aimed to investigate the thermal performance of a hundred (100) NCR18650 lithium cells battery modules in a battery pack with four vents. The test process used in this study aimed to simulate as closely as possible near real-time application scenarios, hence the ambient conditions such as room temperature are considered an uncontrollable parameter so that little or no action is taken to control or alter ambient conditions during the test period.

After defining factors and their levels to be tested for experiment in this study, various design of experiment (DOE) methods such as Plackett–Burman, Taguchi, latin square and full factorial posed as viable methods to be used in planning and designing the experiments to be carried out. After assessing various strength and features of each method, the "full factorial experiment design" was settled for as it allowed for a study of the main and interacting factor (air and current flow rate) effects on the battery module and also allowed for the development of a response surface of the design space tested. The full factorial method employed also provided the maximum number of experiments be performed for selected factors and levels. A total number of thirty-six (36) experiments were carried out (due to repetition) and twelve (12) unique experiments analyzed after the results of experiments with similar combinations of factors were averaged.

Table 3 below shows the various factors and levels tested during the experiment and the experiment design development code using MATLAB (R2016B, Mathworks, Inc., Natick, MA, USA, 2016).

Table 3. Design of Experiment Process.

Factors	Levels
Current Rate (C)	0.5 [1]
	0.75 [2]
Air flow Rate (m/s)	1.4 [1]
	2.4 [2]
	3.4 [3]

```
% full factorial([2 levels of current rate, 3 levels of flow rate])
>> DOE = full fact([2,3])
DOE=
   1   1
   2   1
   1   2
   2   2
   1   3
   2   3
```

Table 3. *Cont.*

Repetition	Experiment Map					
	0.5 C			0.75 C		
	1.4 m/s	2.4 m/s	3.4 m/s	1.4 m/s	2.4 m/s	3.4 m/s
1	Charge	Charge	Charge	Charge	Charge	Charge
	Discharge	Discharge	Discharge	Discharge	Discharge	Discharge
2	Charge	Charge	Charge	Charge	Charge	Charge
	Discharge	Discharge	Discharge	Discharge	Discharge	Discharge
3	Charge	Charge	Charge	Charge	Charge	Charge
	Discharge	Discharge	Discharge	Discharge	Discharge	Discharge

2.3. Experiment Procedure

The charge/discharge method employed while testing the battery pack performance followed the constant current–constant voltage (CC–CV) or Galvanostatic method. Employing this method of charging, the battery module is initially charged at a specified current rate of 0.5 C (16 Amps) from its minimum cutoff voltage of 32 V until it barely reaches its maximum cutoff voltage—typically 41.99 V. At this stage in the charging process, the voltage is held at constant until the current flow rate reaches $0.3 - 0.2$ Amps.

Before each cycle of the experiment, the cooling fans were inspected and set to the required level of air flow rate and measured with an anemometer. As the battery pack design employs two inlet cooling fans, the area of the vents calculated in Table 4, was doubled to determine the total volume of air being pushed into the battery pack at every set cooling fan speed.

After each charge/discharge cycle was completed, time was allowed for the cells in the battery module to rest in order to ensure electrochemical stability [14] before a new cycle commenced. This waiting period also allowed for the entire battery module to reach a uniform cooled temperature. Figure 13 illustrates the systematic steps carried out to perform each experiment cycle.

Table 4. Battery Pack Parameters.

Volume Parameters (m^3)	
Total	0.0075
Partitioned	0.00125
Actual	0.00625
Vent Dimensions	
Diameter of Vent (m)	0.035
Area of vent (m^2)	0.003848
Two inlet vents (m^2)	0.007697
Volume Flow rate at various fan Speeds (m^3/s)	
1.4	0.010776
2.4	0.018473
3.4	0.026169

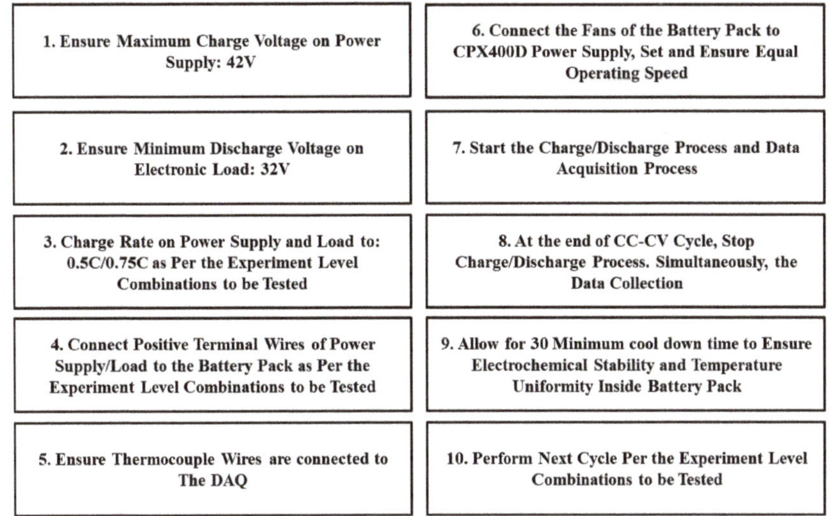

Figure 13. Experiment cycle procedures. CC-CV: constant current–constant voltage.

2.4. Objective Functions Investigated

Temperature profiles of sixteen (16) cells from the battery module were monitored and recorded by the K-type thermocouples and the temperature DAQ as the output variables of the experiment. An average of 172 temperature data points (corresponds to 172 min) were generally recorded during the 0.5 C level charge experiment and 150 data points during the charging experiment with 0.75 C current rate. Generally, lower times of discharging period were observed for the 0.5 C and 0.75 C experiments.

The temperature values recorded during each charge/discharge cycle experiment were stored in a generic created file by the T-DAQ which was retrieved for data processing for data analysis.

For results analysis of the measured temperature profile of selected cells in the battery module, the nomenclature depicted in Figure 14 was adopted to address various individual cells (e.g., Cell#01, Cell#07, etc.), or a group of cells in a row as Row 1, Row 2, Row 3 and Row 4.

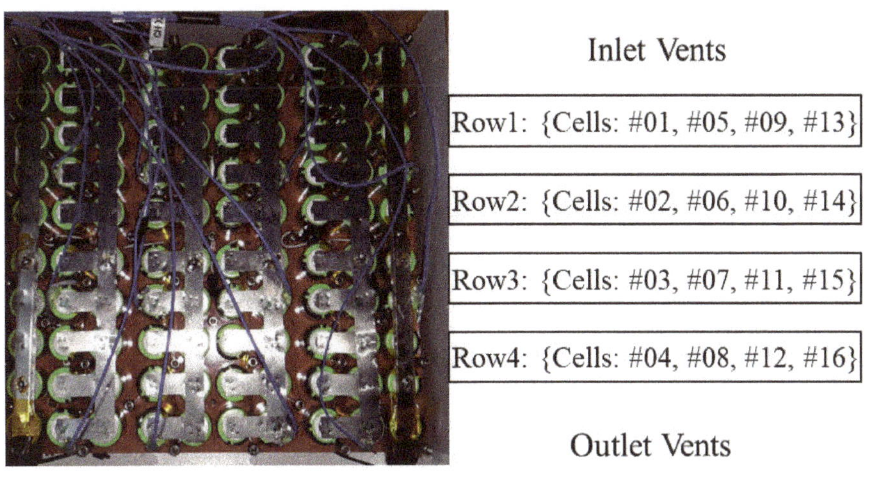

Figure 14. Sensor position nomenclature.

2.4.1. Maximum Temperature (T_{MAX})

The performance and longevity of a cell operating under any given charge or discharge cycle greatly relies on its operating temperature not exceeding 40 °C [13,15,16]. The maximum temperature (T_{MAX}) of any individual cell in a battery pack cooled under any battery thermal management system is therefore indicative of the overall performance of the BTMS system.

2.4.2. Temperature Increase (T_{INC})

The Temperature increase (T_{INC}) represents the temperature difference between the initial temperature (T_I) and the highest temperature (T_H) of each measured cell in the battery pack. This measure, similar to the maximum temperature of a cell in the battery pack, is indicative of the performance of a BTMS but takes into consideration the temperature profile of each measured cell in relation to its neighboring cells. It also allows for the measure of temperature uniformity between cells in similar rows—one (1) through four (4)—or submodules. For a better BTMS performance, temperature uniformity between cells improves charging uniformity.

2.4.3. Temperature Difference (ΔT_{MAX})

The maximum allowable temperature between cells in a BP of 5 °C has been reported in several studies including [16–19], to promote battery balancing and uniform charging and discharging during the LiB's operating cycle. The average temperature of cells in each row was determined to determine the temperature difference among cells in the battery module for each experiment performed.

Typically, an experiment with a combination of design levels which yields a temperature difference among various cells in a battery pack above 5 °C would be considered to have performed poorly.

3. Results and Discussion

This section provides a concise and precise description of the experimental results, their interpretation as well as how they are interpreted within the perspective of previous studies.

3.1. Thermal Performance of BTMS

The maximum temperature experienced by monitored cells in the battery pack during the experiment performed in this study were obtained by averaging the temperature values of experiments performed under similar combinations of design parameters after repetition. Figure 15 presents a capture of the results of all the tests performed in this study.

The data points are plotted based on the arrangements of cells in the battery pack with respect to the cooling-air flow channel; so that maximum temperature (TMAX) of cells in Row 1 (Cells: #01, #05, #9, #13) which are closest to the inlet vents are plotted first following the systematic pattern through to cells in Row 4 based on the illustration presented in Figure 14, page 11 above.

From the graph of results presented in Figure 15 a common trend in the thermal behavior of cells in Row 1, irrespective of the current rate, charging cycle or the cooling-air speed, is that they recorded the least maximum temperature. This can be associated to the fact that they naturally experience the effects of cooling air pumped into the battery pack at its ambient state in terms of temperature and speed. Another factor that plays greatly to this observed trend is the absence of accumulated heat generated by collective cells in the battery pack at the inlet vent area.

Another common trend observed in the results across all the graphs in Figure 15 is the relatively similar maximum temperatures in the battery packs during the discharge cycle and the charge cycle experiments. Quantitatively, the absolute peak temperature obtained by the cell in channel 12 during the 0.5 C charge cycle is 30.7 °C, and 29.3 °C during the discharge cycle under a 1.4 m/s cooling rate. Table 5 presents further comparisons during the charge and discharge cycle for cell #06 observed during tests carried out in this study.

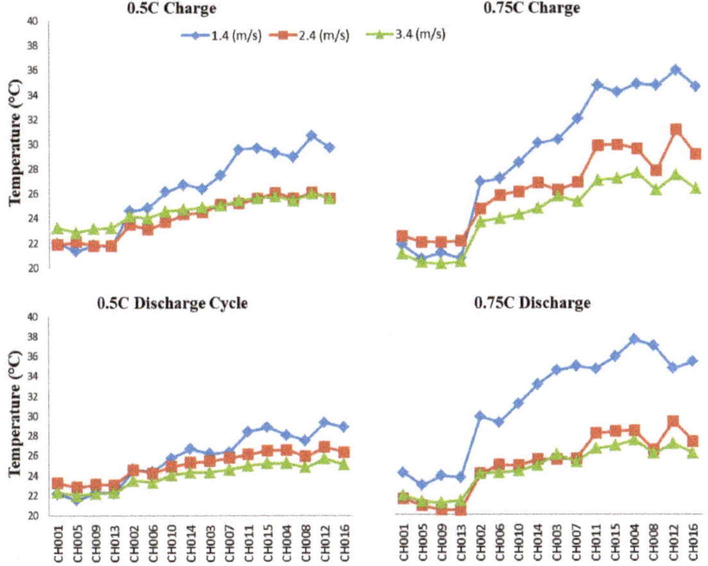

Figure 15. Thermal characteristics of the battery pack.

Table 5. Disparities in Maximum Temperature of Cell 12 during Charge/Discharge Cycle.

	0.5 C		0.75 C	
	Charge	Discharge	Charge	Discharge
1.4 m/s	30.7 °C	29.3 °C	36.1 °C	34.6 °C
2.4 m/s	26.07 °C	26.8 °C	31.1 °C	29.3 °C
3.4 m/s	26.0 °C	25.72 °C	27.7 °C	27.1 °C

In the data presented in Figure 15 there is a relatively bigger difference in the thermal behavior of the battery pack during its operation under a cooling-air flow rate of 1.4 m/s as compared to the performance between the air flow rates of 2.4 m/s and 3.4 m/s. The closeness in performance of the battery pack for air flow rate 2.4 m/s and 3.4 m/s is observed for charging and discharging under 0.5 C and for discharging under 0.75 C while the trend does not hold true for charging under 0.75 C. This break in trend can be associated with the tendency of lithium ion cells to generate significantly heat at a higher current rate.

The general increase in the trend of the maximum temperature registered in the cells as they move further away from the inlet vents is observed for cells 1 through 16 in the entire test performed and can be associated with the effects of heat accumulation and increase in resistance of the flow of the cooling-air path. Similar effects have been reported in [18,20] and measurements such as bidirectional air flow have been proposed and investigated for battery packs minimizing cell maximum temperature and inter cell temperature difference hence improving temperature uniformity in the cells of a battery pack [18,20,21].

A critical study of the thermal result of the battery module presented in Figure 15, cellmonitored by channel 12 of the T-DAQ is noticed to always record slightly lower temperatures than cells in its locality (Row 4). Upon investigating this behavior, it was observed that Cell 12 happens to be directly in front of an exhaust vent hence it is hypothesized that Cell 12 experienced slightly better cooling as the heated air in the battery pack was constantly vented from its position. This phenomenon would also prevent heat accumulation at that locality.

Finally, comparing results obtained for the battery pack performance presented in this study, an averaged maximum temperature of 36.1 °C was obtained for the 0.75 C charge experiment at a 1.4 m/s flow rate and a temperature value of 37.5 °C after three repetitions of discharging at 0.75 C and 1.4 m/s. In a real-life application, charging under such conditions (0.75 C & 1.4 m/s) would not be recommended as the optimal temperature for operating lithium ion cells is 40 °C [13,22]: a value which the worst-obtained results in this study (4 to 5 °C) is just shy of.

3.1.1. Effects of Air Flow Rate on Maximum Temperature

From previous research conducted in the literature review stage of this study, the general trend observed in many published research under BTMS studies is that a higher cooling-air flow rate yields better BTMS performance in terms of the objective functions: minimization of maximum cell temperature, increases temperature uniformity amongst cells and minimizes temperature differences between cells.

A similar trend has been observed for the battery pack model presented and tested in this study. As shown in Figure 16. The test results presented in Figure 16 just as in Figure 15 are obtained after averaging the temperature data recorded for the unique experiments after repetition.

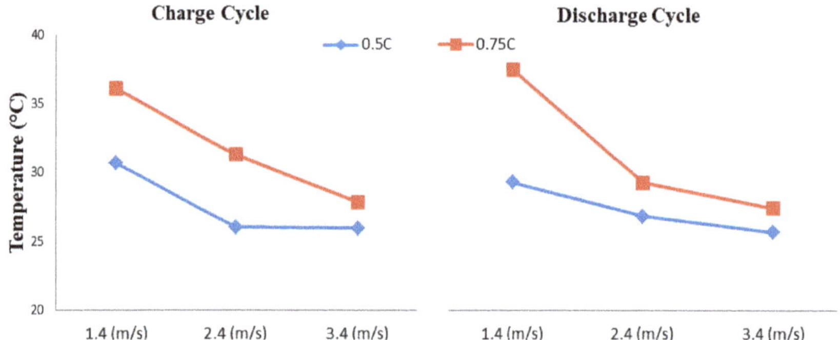

Figure 16. Effects of air increasing air flow rate on maximum temperature.

3.1.2. Effects of Air Flow Rate on Temperature Difference

Investigating the performance of the battery pack for temperature difference between cells, the average temperature of cells in each row (one through four) is determined for all the individual unique experiments. This measure taken reduces the measured temperature output from sixteen cells to four cells classified by their positions in the battery pack (see Figure 14, page 11 above). After classification of all sixteen cells into four rows, the maximum temperature in each row is determined. Lastly, the absolute difference in T_{MAX} between all the combinations of rows is presented in Figures 17 and 18.

In Figure 17, it is observed that for the lower current rate of 0.5 C tested, the majority of temperature differences between rows measured in the battery pack is below 5 °C. The maximum temperature difference measured during experiments conducted under the 0.5 C current rate was recorded to be 8 °C during a charge cycle under 1.4 m/s and 6.37 °C for a discharge cycle under the same flow rate which occurred between cells in Row 1 and Row 4.

For flow rates of 2.4 and 3.4 m/s tests under 0.5 C, the temperature difference between all interacting rows in the BP was kept well below 5 °C.

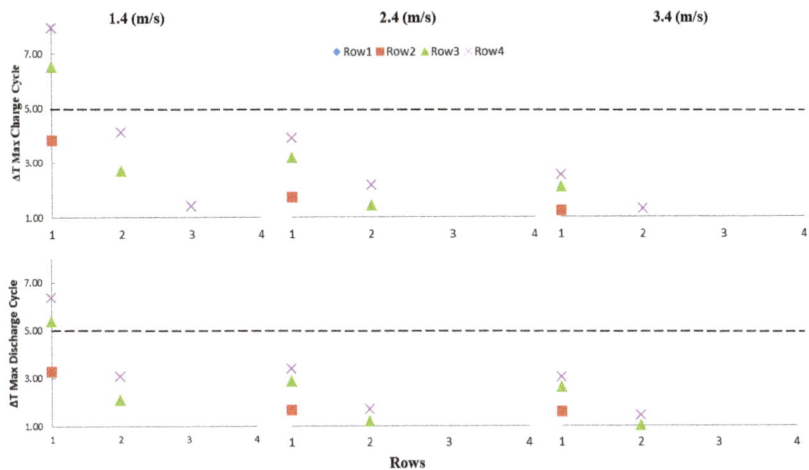

Figure 17. Effects of air flow rate on temperature differences (0.5 C).

In Figure 18, which plots the cell temperature difference for experiments conducted under a 0.75 C current rate, the temperature differences between Row 1 and Row 2, Row 1 and Row 3, Row 1 and Row 4 and Row 2 and Row 4 were observed to be above the 5 °C threshold of under air flow rate of 1.4 m/s during the charge and discharge experiments.

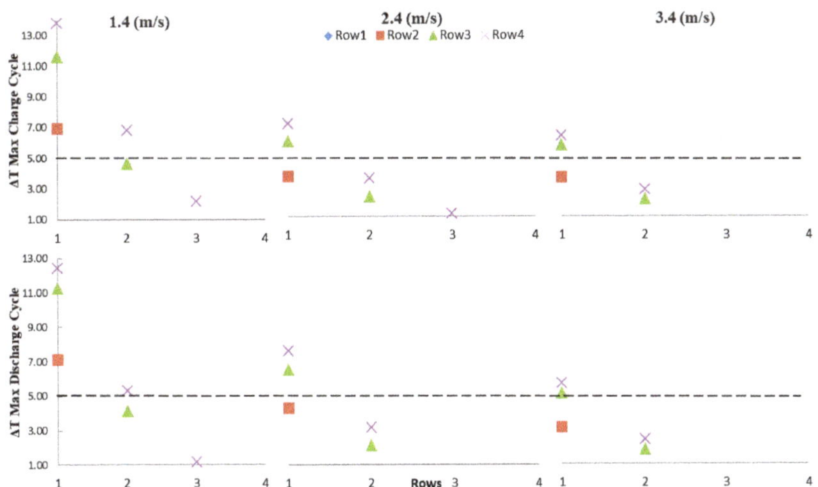

Figure 18. Effects of air flow rate on temperature differences (0.75 C).

As the air flow rate increases to 2.4 m/s, the temperature difference which occurred under the charge cycle is reduced from three to two interactions (Rows 1 and 3, Rows 1 and 4) similar to results obtained for the 0.5 C charge experiment under 1.4 m/s.

The maximum temperature differences between Rows 1 and 4 during the charge and discharge cycle under 0.75 C were 13.81 °C and 12.42 °C respectively for 1.4 m/s flow rate. These values reduced to 6.0 °C and 5.25 °C under an air flow rate of 3.4 m/s.

3.2. Battery Pack Model Development

In this section, two model development techniques are applied to the data set applied and obtained from the experiments in this study. Raw input data as described in Table 6 and maximum temperature data obtained from the 16 monitored cells are used as input and output training data for an artificial neural network (ANN) algorithm to develop a model for the battery pack investigated in this study. All 12 unique current rate and air flow rate combination experiments developed by the full factorial DOE method are used as input data (X,Y) with the averaged absolute maximum temperature of each unique experiment used as an output (Z) to develop a surface regression model.

Table 6. Artificial Neural Network (ANN) Input Training Data Sample.

Experiment	Cycle	Flow Rate (m/s)	Current Rate (C)	Ambient Temp °C
0.5DS1.3	0	1.4	0.5	17.10
0.75DS2.1	0	2.4	0.75	18.76
0.75CS1.1	1	1.4	0.75	19.08

3.2.1. Artificial Neural Network

An artificial neural network (ANN) is a set of interconnected neurons that mimic information processing, similar to humans. It provides a function for creating, training, visualizing and simulating neural networks capable of performing classification, regression, clustering time-series forecasting and dynamic modeling [23,24]. A neural network consists of a two layer feed-forward network with sigmoid hidden neuron and linear output neurons that can fit multidimensional mapping problems arbitrarily well, given consistent data and enough neurons.

Several types of neural network and how they are applied to solve the specific problems they are suited for have been demonstrated in literature. Peculiar to research on lithium ion batteries, X. Qian, et al. in [25] applied neural networks in optimization of his design parameters for a proposed battery pack model and H. Sassi et al. applied ANN to an empirical data set to develop a model to predict the State of Charge (SOC) level of lithium ion cells studied in their work [23].

In this study, an ANN architecture model (Figure 19) is trained offline with the input data set from all experiments conducted in this study. Input data included the charge cycle (represented as 1), discharge cycle (represented as 0), the ambient temperature during the experiment, the current rate and the air flow rate (Table 6) while the maximum temperature of all monitored cells were fed as output data to the neural network.

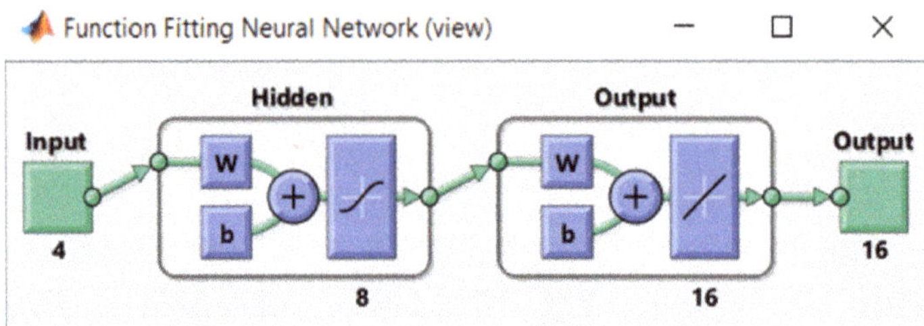

Figure 19. ANN architecture.

In total, a matrix of 4 by 37 and 16 by 37 data size were used as input and output data sets respectively. Seventy percent of input data was allocated for the neural network for model training while 15% each was allocated for the testing and validation phase of the ANN model development.

After several iterations of training, a model obtained with an R value (correlation value) of 0.99139 was obtained between the fitted model and the given training data, 0.9812 correlation R value between the fitted model and the given data for testing and an R value of 0.98869 for an overall correlation between the actual outputs and the targets was obtained (see Figure 20). In Figure 21, the error histogram is plotted showing the difference between the target and actual output of the ANN training, which revealed that among the total samples considered, the majority of the error lies in the range 0.1034 to 2.08.

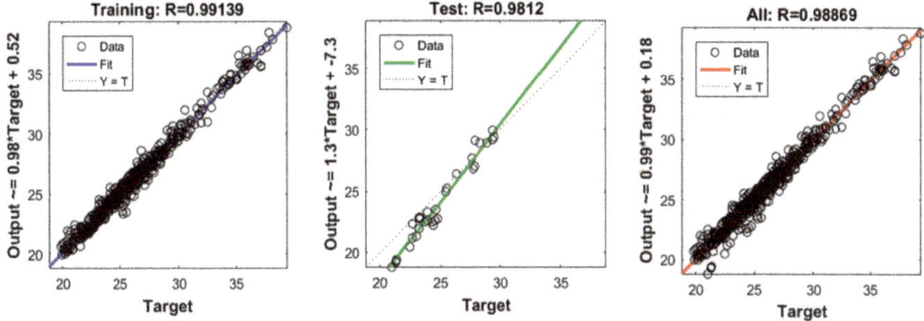

Figure 20. ANN regression plot.

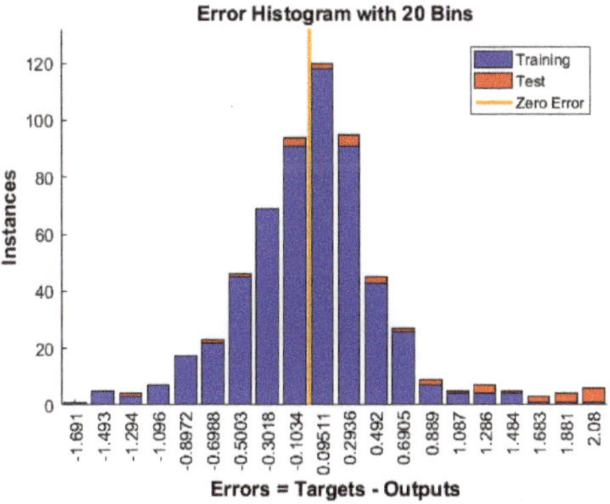

Figure 21. Error histogram.

The training algorithm selected for training the ANN module in this thesis study was the 'Bayesian Regularization' algorithm for its performance with difficult, small and noisy data sets [24]—a critical feature of the data output obtained during the conducted experiments in this study.

3.2.2. ANN Model Validation/Error Analysis

After the ANN model was developed and trained with experimental data, a simulink model illustrated in Figure 22 and a MATLAB function code was generated as a representation of the model. In this chapter, the accuracy of the model developed was tested in a closed loop fashion by comparing the model output value to a set of experimentally obtained output values and finding the absolute percentage error between the two values for a given set of input data. A lower percentage error value between a predicted and measured value is desired for a good model and implies such a model will be good for studying the physical model applying optimization if need be.

Figure 23 shows the maximum temperature comparison between empirically obtained data for a discharge cycle experiment conducted with a 0.75 C current rate and 2.4 m/s air flow rate versus the obtained maximum temperature values predicted by the trained model. Figure 23 shows a close relation between the experiment and predicted model output with an absolute maximum percentage error of 1.84%. As Figure 23 compares and presents the relative error between the outputs of the trained ANN model and actual experimental output values for a given input data set, Figure 24 illustrates the error percentage graph between the predicted and actual maximum temperature values for the entire experiment input data. The data in Figure 24 showed that 93% of the entire input-data-predicted output by the ANN model when compared to their counterpart experimentally obtained output had an absolute percentage error of less than 4%. A maximum percentage error of 10.38% between a predicted and an actual output for a given set of input was measured.

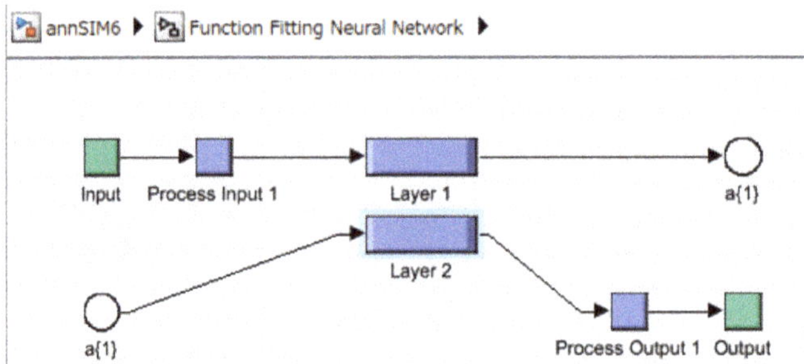

Figure 22. ANN simulink model.

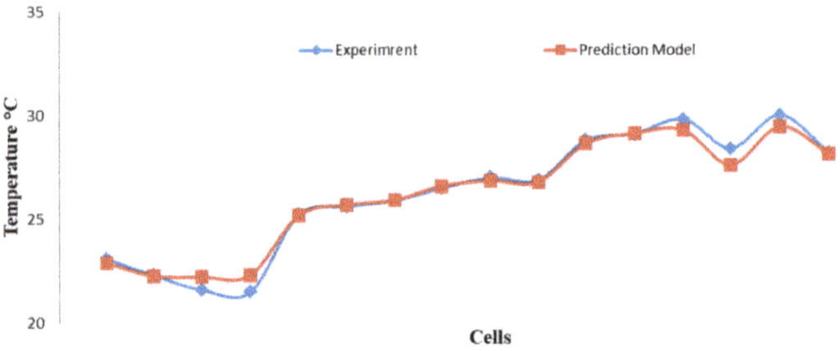

Figure 23. Predicted maximum cell temperature (Tmax) versus actual value.

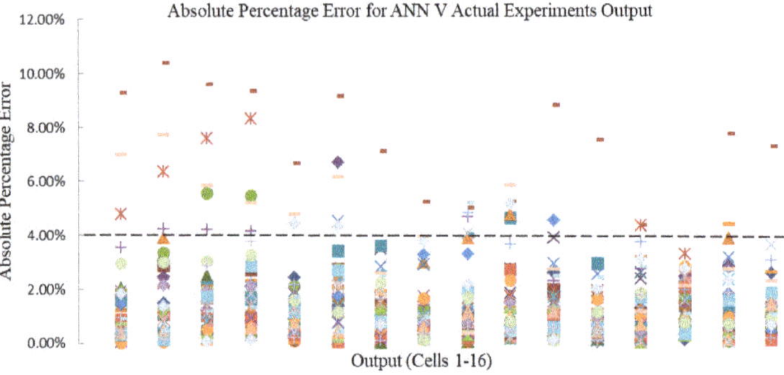

Figure 24. Percentage error between real output and predicted output values for the entire experiments.

3.2.3. Response Surface Model (RSM) Development

An RSM comprises regression surface fitting over a bounded design space to predict approximated responses for input variable combinations not accounted for during a physical or simulation experiment [26]. Usually, a design of experiment method is used to obtain the minimum number of experiments needed to develop an RSM. In the RSM method implemented in this study, the approximation function used in developing the response surface illustrated in Figure 25 is a second degree polynomial. Most response surfaces functions are generated with polynomials depending on the number of data points provided from an experiment.

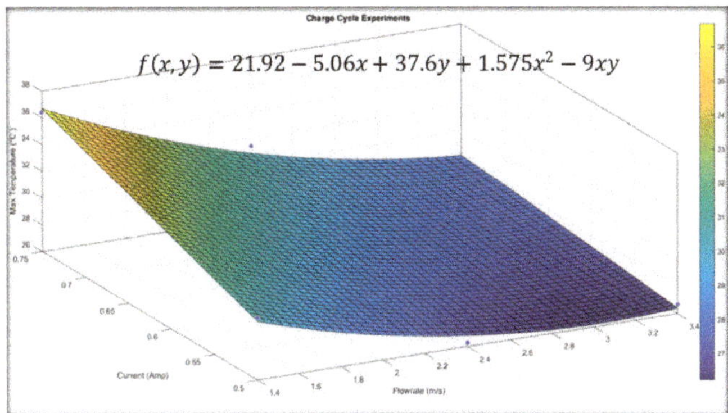

Goodness of Fit: SSE- 0.9075, R^2- 0.9875, Adjusted R^2- 0.9373

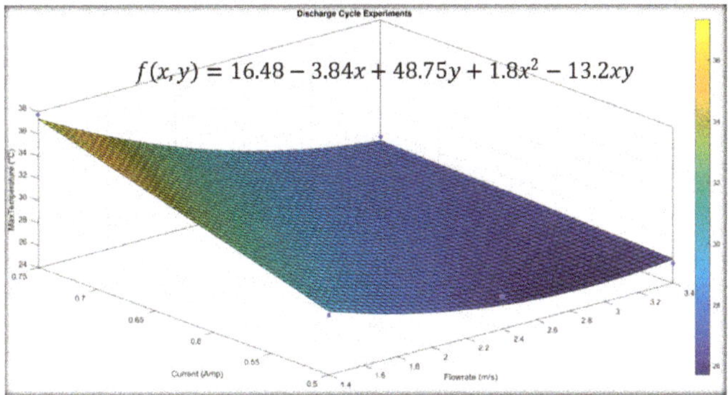

Goodness of Fit: SSE- 1.6133, R^2- 0.9824, Adjusted R^2- 0.9121

Figure 25. Response surface model (RSM) developed for charge and discharge experiment.

The RSM developed for the battery model was done using a MATLAB curve fitting tool box. Fitting parameters R^2 adjusted were used to determine the goodness of fit, and sum of square error (SSE) and mean square error are used to determine the predictability of the model. The R^2 adjusted values lie between 0 and 1 for which a good response surface has a value closest to 1 [26,27]. On the contrary, an SSE and Mean Square Error (MSE) value closer to zero is desired for a good response surface. An SSE value closer to zero indicates a model has smaller random error components hence higher prediction accuracy [27]. Just as for SSE, an MSE value closest to zero is desired for a good response model.

After various variations of the fit method applied on the variables were tested to obtain the best adjusted R^2 and SSE values, the model was developed with the polynomial function with robustness "off" as this trail produced the most suitable fitness parameters results. Table 7 provides a comparison of various selection criteria values obtained during the iterative training process.

Table 7. Comparison of Various Fitness Parameters for the Developed Regression Model.

	2nd Degree Polynomial Fit				
Robustness:			Off		
	Fit Type	SSE	* RMSE	R^2	Adjusted R^2
Charge	** poly21	0.9075	0.9526	0.9875	0.9373
Discharge	poly21	1.6133	1.2702	0.9824	0.9121
Robustness:			Least Absolute Residual (LAR)		
Charge	poly21	0.9075	0.9526	0.9875	0.9373
Discharge	poly21	1.6133	1.2702	0.9824	0.9121
Robustness:			Bi-Square		
Charge	poly21	1.3491	1.1615	0.9814	0.9068
Discharge	poly21	2.3984	1.5487	0.9739	0.8693

* Root Mean Square Error. ** poly21: A second degree polynomial fit with two degrees of X and one degree of Y. SSE: sum of square error.

The model equation of the RSM developed for the charge and discharge experiments performed for the battery pack and module in this study is presented as:

$$f(x, y) = P_0 + P_1 x + P_2 y + P_3 x^2 + P_4 xy \tag{1}$$

where x and y are designed variables air flow rate and current rate respectively, $P_0 - P_4$ are coefficient constants and $f(x, y)$ is the design objective function. Table 8 presents coefficient values for the developed model equations.

Table 8. Model Equation Coefficients.

	Charge	**Discharge**
P_0	21.92	16.48
P_1	−5.06	−3.84
P_2	37.6	48.75
P_3	1.575	1.8
P_4	−9	−13.2
	Goodness of Fit	
Charge:	SSE—0.9075, R^2—0.9875, Adjusted R^2—0.9373	
Discharge:	SSE—1.6133, R^2—0.9824, Adjusted R^2—0.9121	

3.2.4. RSM Model Validation/Error Analysis

In an attempt to validate the regression model developed in this study, equations of the developed RSM's were tested for experimental inputs to compare the output results. Figure 26 plots and compares the predicted absolute maximum temperatures against the actual maximum temperatures for the charge experiment at 0.5 C and the discharge experiment at 0.75 C.

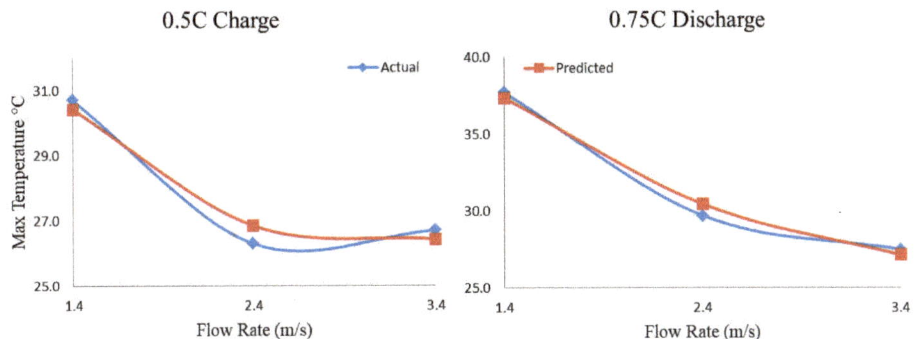

Figure 26. Regression model predicted output vs. actual output for a sample input.

Table 9 below shows the absolute relative error calculations between the predicted model and the actual outputs for absolute maximum temperature in a battery pack for all twelve of the unique experiments performed. The results showed the developed model predicts accurately with an absolute maximum error of approximately 3.00%.

Table 9. Absolute Relative Error between Regression Model and Actual Experiment.

	Charge at 0.5 C			Discharge at 0.5 C		
	1.4 (m/s)	2.4 (m/s)	3.4 (m/s)	1.4 (m/s)	2.4 (m/s)	3.4 (m/s)
Actual (°C)	30.70	26.30	26.70	29.40	26.90	25.80
Predicted (°C)	30.42	26.85	26.42	29.77	26.17	26.17
% Error	0.91%	2.08%	1.04%	1.25%	2.72%	1.42%
	Charge at 0.75 C			Discharge at 0.75 C		
	1.4 (m/s)	2.4 (m/s)	3.4 (m/s)	1.4 (m/s)	2.4 (m/s)	3.4 (m/s)
Actual (°C)	36.40	31.40	27.90	37.70	29.70	27.50
Predicted (°C)	36.67	30.85	28.17	37.33	30.43	27.13
% Error	0.75%	1.76%	0.98%	0.97%	2.47%	1.33%

4. Conclusions

In this study, a battery thermal management system (BTMS) for a battery pack housing a battery module consisting of a hundred NCR18650 lithium ion cells was designed and tested in relatively cold ambient conditions.

The BTMS performance was tested for two major objective function criteria which are:

- maximum temperature recorded by a cell in the battery pack which at any instant should be kept below 40 °C and
- maximum temperature difference between any cells in the battery pack during testing which should not exceed a threshold of 5 °C.

The temperature limit threshold for operating LiBs was optimally obtained from various literature and battery data specification documents.

Design variables—three levels of cooling-air flow rate and two levels of current rate were combined using a full factorial experiment design method to develop a full array of experiments performed on the BTMS for the battery pack after experimentation. General BTMS performance trends were observed in the obtained results such as higher current rate experiments produced relatively higher maximum temperature amongst the cells and significant changes in maximum cell temperature and temperature difference are observed as air flow rate is increased.

Upon investigating the effects of increasing air flow rate on maximum cell temperature, a 15.09% reduction in maximum temperature recorded by a cell was achieved during a charging experiment with a 0.5 C current rate by increasing the air flow rate from 1.4 m/s to 2.4 m/s. For the same charge experiment, there was no significant improvement in the maximum temperature recorded at a higher air flow rate of 3.4 m/s. On the contrary, at a higher current rate of 0.75 C charging, a 13.20% reduction in the maximum temperature of a cell was achieved by increasing the air flow rate from 1.4 m/s to 2.4 m/s. Further increment of the air flow rate to 3.4 m/s produced a 22.81% reduction in the maximum cell temperature.

The results summarized above aids in drawing a hypothesis that for the investigated battery pack design, for each current rate, there exists an optimal cooling-air flow rate that if exceeded, will yield little or no improvement in an operating BTMS performance at a specific range of ambient conditions. This hypothesis will prove vital in scenarios where power consumption of an operating BTMS is a critical objective function to be minimized. This will be applicable and vital to the development of an intelligent/dynamic BTMS where cooling operation of a BTMS will operate in a dynamic mode depending on the current profile the battery module.

When assessing the BTMS performance based on temperature difference between cells, for a cooling-air flow rate of 1.4 m/s for every cycle experiment, there was always an instant where the maximum temperature difference between monitored cells exceeded 5 °C, reducing the performance of the BTMS. However, at higher speeds of 2.4 m/s and 3.4 m/s, in experiments under 0.5 C, the temperature difference among cells in the battery pack was found to be always below the 5 °C threshold.

Lastly, for 0.75 C experiments under 1.4 m/s, 100% of the interactions between cells in Rows 2, 3 and 4 with Row 1 exceeded 5 °C. However, after increasing the air flow rate temperature difference between cells in Rows 2, 3 and 4 with Row 1 which exceeded 5 °C, is reduced by 33.33%. The highest temperature recorded under 0.75 C with 2.4 m/s was found to be 7.71 °C (a 44.14% reduction from operation under 1.4/s) and 6.3 °C (a 54.38% reduction from operation under 1.4/s) for 3.4 m/s between cells in Row 4 and Row 1.

The observations made aides in affirming the conclusion based on literature that a higher cooling-air flow rate reduces maximum cell temperature and temperature difference between cells in a battery pack due to the increased convective heat transfer coefficient of air at higher velocities. However, the gradient in temperature difference amongst cells remains the same with at least a single case of interaction between cells at different positions in the battery module exceeding 5 °C for the investigated battery module.

To fully obtain an air-cooled BTMS performance with the possibility of no interacting cells in the battery module having ΔTmax exceeding 5 °C, a bidirectional cooling flow path scheme must be implemented.

Author Contributions: The work carried out and presented in this manuscript was carried out as part of A.A.A.H. Master's program completion requirement under the close supervision of D.S. as his thesis project supervisor. All authors have read and agreed to the published version of the manuscript.

Funding: This research received no external funding.

Acknowledgments: The authors would like to acknowledge the entire EMU EVDC [28] and the director Davut Solyali for the support provided in the successful completion of this project in terms of experimental equipment setup and a conducive laboratory working environment.

Conflicts of Interest: The authors declare no conflict of interest.

References

1. Kim, J.; Oh, J.; Lee, H. Review on battery thermal management system for electric vehicles. *Appl. Therm. Eng.* **2019**, *149*, 192–212. [CrossRef]

2. IEA. *Global EV Outlook 2017 Together Secure Sustainable Global EV Outlook 2017*; International Energy Agency: Paris, France, 2017; pp. 5–55.

3. Wu, W.; Wang, S.; Wu, W.; Chen, K.; Hong, S.; Lai, Y. A critical review of battery thermal performance and liquid based battery thermal management. *Energy Convers. Manag.* **2019**, *182*, 262–281. [CrossRef]

4. Akinlabi, A.A.H.; Solyali, D. Configuration, design, and optimization of air-cooled battery thermal management system for electric vehicles: A review. In *Renewable and Sustainable Energy Reviews*; Elsevier Ltd.: Amsterdam, The Netherlands, 2020; p. 109815.

5. Budde-Meiwes, H.; Drillkens, J.; Lunz, B.; Muennix, J.; Rothgang, S.; Kowal, J.; Sauer, D. A review of current automotive battery technology and future prospects. *Proc. Inst. Mech. Eng. Part D J. Automob. Eng.* **2013**, *227*, 761–776. [CrossRef]

6. Li, L.; Dababneh, F.; Zhao, J. Cost-effective supply chain for electric vehicle battery remanufacturing. *Appl. Energy* **2018**, *226*, 277–286. [CrossRef]

7. Adolf, J.; Balzer, C.H.; Louis, J.; Schabla, U.; Fischedick, M.; Arnold, K.; Pastowski, A.; Schüwer, D. *Shell Hydrogen Study Energy of the Future? Sustainable Mobility through Fuel Cells and H2*; Technical Report; Shell Deutschland Oil GmbH: Hamburg, Germany, 2017.

8. Arora, S. Selection of thermal management system for modular battery packs of electric vehicles: A review of existing and emerging technologies. *J. Power Sources* **2018**, *400*, 621–640. [CrossRef]

9. Chen, K.; Wang, S.; Song, M.; Chen, L. Configuration optimization of battery pack in parallel air-cooled battery thermal management system using an optimization strategy. *Appl. Therm. Eng.* **2017**, *123*, 177–186. [CrossRef]

10. Yang, N.; Zhang, X.; Li, G.; Hua, D. Assessment of the forced air-cooling performance for cylindrical lithium-ion battery packs: A comparative analysis between aligned and staggered cell arrangements. *Appl. Therm. Eng.* **2015**, *80*, 55–65. [CrossRef]

11. Lu, Z.; Yu, X.; Wei, L.; Qiu, Y.; Zhang, L.; Meng, X.; Jin, L.W. Parametric study of forced air cooling strategy for lithium-ion battery pack with staggered arrangement. *Appl. Therm. Eng.* **2018**, *136*, 28–40. [CrossRef]

12. Chen, K.; Chen, Y.; She, Y.; Song, M.; Wang, S.; Chen, L. Construction of effective symmetrical air-cooled system for battery thermal management. *Appl. Therm. Eng.* **2020**, *166*, 114679. [CrossRef]

13. Panasonic. *Datasheet: Specifications Powercell NCR18650B*; SANYO Energy Corporation: San Diego, CA, USA, 2012; No. 469; p. 1.

14. Gümüşsu, E. Thermal Modeling of Lithium Ion Batteries Lityum Iyon Pillerin Isi L Modellemesi. Master's Thesis, Hacettepe University, Ankara, Turkey, 2017.

15. Al-Zareer, M.; Dincer, I.; Rosen, M.A. Performance assessment of a new hydrogen cooled prismatic battery pack arrangement for hydrogen hybrid electric vehicles. *Energy Convers. Manag.* **2018**, *173*, 303–319. [CrossRef]

16. Li, K.; Yan, J.; Chen, H.; Wang, Q. Water cooling based strategy for lithium ion battery pack dynamic cycling for thermal management system. *Appl. Therm. Eng.* **2018**, *132*, 575–585. [CrossRef]

17. E, J.; Han, D.; Qiu, A.; Zhu, H.; Deng, Y.; Chen, J.; Zhao, X.; Zuo, W.; Wang, H.; Chen, J.; et al. Orthogonal experimental design of liquid-cooling structure on the cooling effect of a liquid-cooled battery thermal management system. *Appl. Therm. Eng.* **2018**, *132*, 508–520. [CrossRef]

18. Wang, S.; Li, K.; Tian, Y.; Wang, J.; Wu, Y.; Ji, S. Improved thermal performance of a large laminated lithium-ion power battery by reciprocating air flow. *Appl. Therm. Eng.* **2019**, *152*, 445–454. [CrossRef]

19. Pesaran, A. Battery thermal models for hybrid vehicle simulations. *J. Power Sources* **2002**, *110*, 377–382. [CrossRef]

20. Yu, K.; Yang, X.; Cheng, Y.; Li, C. Thermal analysis and two-directional air flow thermal management for lithium-ion battery pack. *J. Power Sources* **2014**, *270*, 193–200. [CrossRef]

21. Lu, Z.; Meng, X.; Wei, L.; Hu, W.; Zhang, L.; Jin, L.W. Thermal Management of Densely-packed EV Battery with Forced Air Cooling Strategies. *Energy Procedia* **2016**, *88*, 682–688. [CrossRef]

22. Liu, X.; Chen, Z.; Zhang, C.; Wu, J. A novel temperature-compensated model for power Li-ion batteries with dual-particle-filter state of charge estimation. *Appl. Energy* **2014**, *123*, 263–272. [CrossRef]

23. Sassi, H.B.; Errahimi, F.; Es-Sbai, N.; Alaoui, C. A comparative study of ANN and Kalman Filtering-based observer for SOC estimation. *IOP Conf. Ser. Earth Environ. Sci.* **2018**, *161*, 12022. [CrossRef]

24. View Neural Network—MATLAB View. Available online: https://www.mathworks.com/help/deeplearning/ref/view.html?s_tid=doc_ta (accessed on 28 December 2019).

25. Qian, X.; Xuan, D.; Zhao, X.; Shi, Z. Heat dissipation optimization of lithium-ion battery pack based on neural networks. *Appl. Therm. Eng.* **2019**, *162*, 114289. [CrossRef]
26. Li, Z.-Z.; Cheng, T.-H.; Xuan, D.-J.; Ren, M.; Shen, G.-Y.; Shen, Y.-D. Optimal design for cooling system of batteries using DOE and RSM. *Int. J. Precis. Eng. Manuf.* **2012**, *13*, 1641–1645. [CrossRef]
27. Goodness of Fit Statistics. Available online: https://web.maths.unsw.edu.au/~{}adelle/Garvan/Assays/GoodnessOfFit.html (accessed on 21 August 2019).
28. EVDC—Electric Vehicle Development Center—EMU. Available online: https://evdc.emu.edu.tr/en (accessed on 21 August 2019).

Article

Torque and Battery Distribution Strategy for Saving Energy of an Electric Vehicle with Three Traction Motors

Yi-Hsiang Tseng [1] and Yee-Pien Yang [1,2,*]

[1] Department of Mechanical Engineering, National Taiwan University, Taipei 10617, Taiwan; gn0000023@gmail.com

[2] Mechanical and Mechatronics Systems Research Laboratories, Industrial Technology Research Institute, Hsinchu 31057, Taiwan

* Correspondence: ypyang@ntu.edu.tw or yeeyang@itri.org.tw; Tel.: +886-2-3366-2682

Received: 12 March 2020; Accepted: 8 April 2020; Published: 11 April 2020

Abstract: A torque and battery distribution (TBD) strategy is proposed for saving energy for an electric vehicle (EV) that is driven by three traction motors. Each traction motor is driven by an independent inverter and a battery pack. When the vehicle is accelerating or cruising, its vehicle control unit determines the optimal torque distribution of the three motors by particle swarm optimization (PSO) theory to minimize energy consumption on the basis of their torque–speed–efficiency maps. Simultaneously, the states of charge (SOC) of the three battery packs are controlled in balance for improving the driving range and for avoiding unexpected battery depletion. The proposed TBD strategy can increase 7.7% driving range in the circular New European Driving Cycle (NEDC) of radius 100 m and 28% in the straight-line NEDC. All the battery energy can be effectively distributed and utilized for extending the driving range with an improved energy consumption efficiency.

Keywords: torque and battery distribution; particle swarm optimization; electric vehicle

---

1. Introduction

Hybrid and pure electric vehicles (EVs) have been commercially available for many years. A lot of research that focused on multiple propulsion and energy storage systems and the related power split and energy economy strategies has become imperative issue in EVs. For EVs with only one battery pack, it is important to balance the capacity of battery cells for improving its lifetime. Li et al. [1] presented a real-time state-of-charge (SOC) calculation method for a pure EV, where the lithium battery was simulated with a second-order resistance–capacitance (RC) model and the remaining capacity in battery cells was balanced by fuzzy control through a set of bi-directional fly-back direct current-direct current (DC–DC) converters. Gallardo-Lozano et al. [2] introduced a shunting transistor method to balance battery cells during the recharging and driving modes. Huang and Abu Qahouq [3] proposed an energy sharing control scheme to regulate the DC bus voltage, and simultaneously, to balance the SOC of battery cells with micro DC–DC converters. Pham et al. [4] addressed a fast-balancing topology for lithium-ion batteries in an EV by transferring the power in high-voltage cells directly to low-voltage cells through DC–DC converters. During battery charging, Dung et al. [5] eliminated the racing phenomenon by a pulse width modulation (PWM) based equalization process among battery packs so that the charging time was reduced by 48%.

For EVs with hybrid energy storage systems (HESSs), Jin et al. [6] and Akar et al. [7], respectively, proposed for their HESS of batteries and ultracapacitors a fuzzy control-based power management strategy to reduce battery degradation. A PWM technique was introduced by Menon et al. [8] for balancing the SOC of independent battery packs by continuously regulating the power flow from two inverters on the basis of the driving demands. Tanaka et al. [9] used two batteries in a hybrid EV for investigating a high-efficiency energy conversion system to improve the driving range. The main battery provided fundamental power that did not need to be passed through a DC–DC converter, while additional power was supplied by a sub-battery through the DC–DC converter.

Lately, pure EVs with multiple traction motors have been commercially available. Examples are the Porsche Mission E Cross Turismo with two permanent magnet synchronous motors (PMSMs) and the Audi e-tron quattro with three traction motors. Rossi et al. [10] introduced a two-motor, two-axle, two-battery pack powertrain configuration for a compact EV and proposed an optimal front-rear motor transmission combination for the best driving performance. Several advantages of multiple motors and battery packs were addressed: the increased fault tolerance; the reduction of power rating in electric drive with possible simplification and cost reduction; the reduction of insulation level and electromagnetic emission of low-voltage power modules; and the additional degrees of freedom in torque vectoring for stable vehicle maneuverability.

Some studies have focused on the driving and braking torque distributions on motors for vehicle stability and handling performance. Yin at al. [11] used a hierarchical electronic stability controller (ESC) to distribute direct torque to four in-wheel motors of an EV for improving the vehicle stability and handling performance. Zhai et al. [12] proposed a similar ESC algorithm to improve vehicle stability by distributing the driving and regenerative braking torque for an EV with four independent in-wheel motors. Other studies have focused on the energy economy of EVs that use torque split strategies to arrange multiple traction motors. Dizqah et al. [13] formulated a parametric energy-efficient torque distribution optimization problem depending on the speed of an EV driven by four identical drivetrains, resulting in an energy consumption reduction of 0.1%–0.5% under various European driving cycles. An EV with four in-wheel motors was introduced by Fujimoto and Harada [14] where the slip ratio and motor loss were optimized on the basis of the vehicle speed and acceleration over the Japanese JC08 cycle. Sun et al. [15] proposed an online braking torque allocation scheme for a four-wheel-drive EV that minimized tire and electromechanical losses. Simulations showed that the driving efficiency was increased 4.3% in high speed driving cycles and 1.5% in normal speed driving cycles.

Yang et al. [16] proposed a real-time torque distribution strategy for a pure EV with three motors and three battery packs. The front wheels were driven indirectly by a traction motor through reduction gears, while two rear wheels were driven directly by two in-wheel motors. Torque distribution was determined by the particle swarm optimization (PSO) theory for minimizing energy consumption on the basis of the torque–speed–efficiency (TNE) maps of all the traction motors. Subsequently, Yang and Chen [17] introduced a coupled parallel energy saving and safety strategy that minimized energy consumption by torque distribution according to the PSO theory. The stabilizing direct yaw moment was also minimized on the basis of the stability region on the phase plane of sideslip angle and yaw rate.

Most of the above research focused either on the battery energy distribution to keep the battery cells in balance, or on the torque distribution for vehicle stability, handling, or energy economy. This paper extends the authors' previous study [18] that proposed a coupled parallel energy balancing and energy saving strategy by keeping the SOC of three independent battery packs in balance and distributing the driving torque of three traction motors during vehicle motion. Section 2 introduces vehicle configuration, longitudinal and lateral vehicle dynamics models, tire and transmission models, and battery SOC model. Section 3 elaborates the proposed torque and battery distribution strategy, and Section 4 provides experiments of model-in-the-loop (MIL) and hardware-in-the-loop (HIL) simulations, and road tests. Section 5 presents concluding remarks.

2. Vehicle Configuration

The EV was fitted with a 15-kW radial-flux PMSM that drove the two front wheels indirectly through a gearbox reducer and two identical 7-kW axial-flux PMSMs to drive the left and right rear wheels directly through the hubs (Figure 1a). The control strategy was validated by CarSim simulation software with 15 mechanical degrees of freedom (DOF) for the four-wheeled vehicle. The steering system had one DOF, each wheel had one spin DOF, each suspension had two DOF, and the sprung mass was simplified as a rigid body with six DOF. The vehicle variables for the longitudinal and lateral dynamics models are defined in Figure 1b,c.

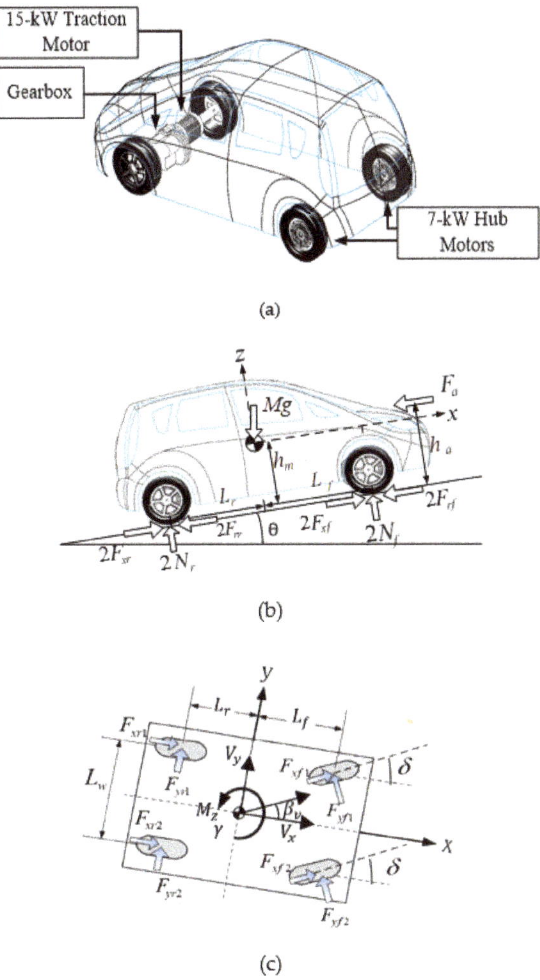

Figure 1. Electric vehicle configuration and variable definitions (**a**) propulsion system of multiple motors, and (**b**) the longitudinal and (**c**) lateral vehicle dynamics models.

Figure 2a,b provide the TNE maps from the experiments for the driving modes of the three traction motors. The braking modes were estimated from the mirror image of 75% efficiency of the driving mode. The maximum torque was 150 Nm and the maximum speed was 2400 rpm for the 15-kW front motor, and they were 122 Nm and 1200 rpm for the 7-kW rear motors. Three motor control units were responsible for driving the three motors, and three lithium-ion batteries were deployed for providing the power: one pack of 144 V, 72 Ah, and 10.45 kWh for the front drive and two packs of 72 V, 72 Ah, and 5.2 kWh for the two rear drives.

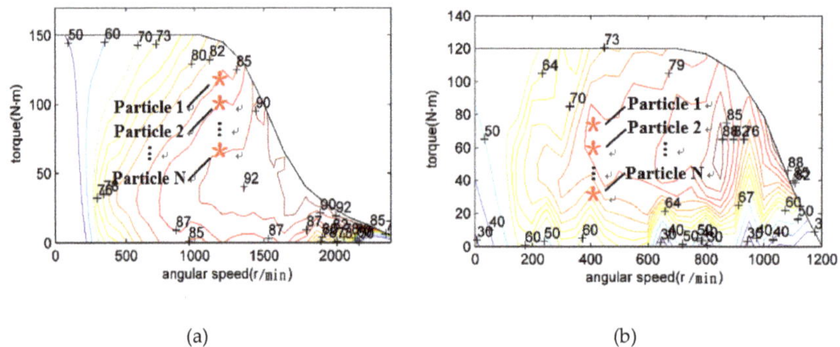

Figure 2. Torque-speed-efficiency maps: (**a**) the driving mode of the 15-kW front motor, (**b**) the driving modes of the two 7-kW rear motors.

2.1. Longitudinal Vehicle Dynamics Model

When the vehicle travels in a straight line, the longitudinal traction forces are usually simplified as $F_{xr} = F_{xr1} = F_{xr2}$ and $F_{xf} = F_{xf1} = F_{xf2}$. The force with subscript 1 is the force exerted on the left tire, while 2 on the right tire. The tractive force F_x and the normal forces N_f and N_r of the front and rear wheels were obtained by the following equations:

$$F_x = M\dot{V}_x + Mg\sin\theta + \frac{1}{2}\rho C_d A_f V_x^2 + 2C_t\left(N_f + N_r\right) \tag{1}$$

$$N_f = \frac{M_v g L_r \cos\theta - \frac{1}{2}\rho C_d A_f V_x^2 h_a - \left[M_v h_m + 2\left(m_{wf} + m_{wr}\right)r_t\right]\dot{V}_x}{2\left(L_f + L_r\right)} + m_{wf}g\cos\theta \tag{2}$$

$$N_r = \frac{M_v g L_f \cos\theta + \frac{1}{2}\rho C_d A_f V_x^2 h_a + \left[M_v h_m + 2\left(m_{wf} + m_{wr}\right)r_t\right]\dot{V}_x}{2\left(L_f + L_r\right)} + m_{wr}g\cos\theta \tag{3}$$

$$F_x = 2\left(F_{xf} + F_{xr}\right) \tag{4}$$

$$M = M_v + 2\left(m_{wf} + m_{wr}\right) \tag{5}$$

where M is the total vehicle mass; M_v is the vehicle mass excluding tire and wheel mass; m_{wf} is the front tire and wheel mass; m_{wr} is the rear tire and wheel mass; g is the gravity acceleration; θ is the slope angle in degrees; ρ is the air density; V_x is the longitudinal velocity of vehicle; F_{xf} and F_{xr} are the traction forces exerted on the front and rear tires; L_f is the distance from mass center to front tire; L_r is the distance from mass center to rear tire; and r_t is the tire radius. Other vehicle specifications used in this paper are described in Table 1.

2.2. Lateral Vehicle Dynamics Model

As shown in Figure 1b, the lateral vehicle dynamics are described by a four-wheel model when the vehicle is cornering. The longitudinal, lateral, and yaw vehicle dynamic equations are expressed as:

$$\left(F_{xf1} + F_{xf2}\right)\cos\delta - \left(F_{yf1} + F_{yf2}\right)\sin\delta + F_{xr1} + F_{xr2} = M\left[\dot{V}_x - \left(\gamma + \frac{d\beta_v}{dt}\right)V_y\right] \tag{6}$$

$$\left(F_{xf1} + F_{xf2}\right)\sin\delta + \left(F_{yf1} + F_{yf2}\right)\cos\delta + F_{yr1} + F_{yr2} = M\left[\dot{V}_y + \left(\gamma + \frac{d\beta_v}{dt}\right)V_x\right] \tag{7}$$

$$L_f(F_{yf1} + F_{yf2}) \cos \delta - L_r(F_{yr1} + F_{yr2}) + \frac{L_w}{2}\left(F_{yf1} - F_{yf2}\right) \sin \delta + M_z = I_z \frac{d}{dt}\left(\gamma + \frac{d\beta_v}{dt}\right) \tag{8}$$

$$M_z = L_f(F_{xf1} + F_{xf2}) \sin \delta + \frac{L_w}{2}(-F_{xf1} + F_{xf2}) \cos \delta + \frac{L_w}{2}(-F_{xr1} + F_{xr2}) \tag{9}$$

where δ is the steer angle; F_{xf1} and F_{xf2} are longitudinal traction forces on the left and right front tires and these forces are assumed equal for $\delta = 0$; F_{yf1} and F_{yf2} are lateral traction forces on the left and right front tires; F_{xr1} and F_{xr2} are longitudinal traction forces on the left and right rear tires; F_{yr1} and F_{yr2} are lateral traction forces on the left and right rear tires; γ is the yaw velocity; β_v is the vehicle sideslip angle; I_z is the mass moment of inertia in the yaw direction; and M_z is the yaw moment for cornering. These equations were used to determine torque distributions when the vehicle is cornering.

Table 1. Vehicle Specifications.

Vehicle Property	Symbol	Value
Frontal area of vehicle [m²]	A_f	1.6
Aerodynamic coefficient	C_d	0.28
Cornering stiffness of the front tire [N/rad]	C_f	51,091
Cornering stiffness of the rear tires [N/rad]	C_r	72,802
Rolling resistance between tire and ground	C_t	0.01
Height of equivalent aerodynamic point [m]	h_a	1
Height of mass center [m]	h_m	0.56
Yaw inertia of vehicle [kg·m²]	I_z	1200
Distance from mass center to front tire [m]	L_f	1.433
Distance from mass center to rear tire [m]	L_r	1.067
Distance between two rear wheels [m]	L_w	1.46
Total mass of vehicle [kg]	M	1813
Sprung mass of vehicle [kg]	M_s	1753
Gear ratio	n_g	3
Tire radius [m]	r_t	0.288

For simplicity, the hill climbing resistance, aerodynamic drag, and rolling resistance of the last terms in (1) are omitted, and the roll and pitch motions are neglected. However, the vehicle in CarSim for the real-time HIL simulation was modelled with self-contained yaw, roll, and pitch dynamics.

2.3. Tire Model

The CarSim tire lookup table that was obtained directly from the laboratory measurements was used to model the tire characteristics. The friction coefficient between the tire and road surface was chosen at 0.85, the longitudinal and lateral traction (or friction) forces on the tire were expressed as a function of normal force and tire slip ratio. Once the vehicle velocity, acceleration or deceleration, the normal forces N_f and N_r are known, the slip ratio of each wheel can be obtained. The front and rear wheel speeds ω_f and ω_r are therefore determined by the definition of tire slip ratio, as follows:

$$Acceleration: \ \omega = \frac{V_x}{r_t(1 - \lambda)}, r_t\omega > V_x \tag{10}$$

$$Deceleration: \ \omega = \frac{V_x(1 + \lambda)}{r_t}, r_t\omega < V_x \tag{11}$$

where the tire slip ratio λ represents λ_f and λ_r that correspond to the speeds ω_f and ω_r of the front and rear wheels.

2.4. Transmission Model

After the wheel speeds and accelerations are obtained, the output torque T_{mf} of the front traction motor, the output torque T_{mr} provided by the right rear in-wheel motor, and the output torque T_{ml} provided by the left rear in-wheel motor can be calculated under the following assumptions:

1. The rotor mass is so small compared with the vehicle mass that the rotational inertias of the three traction motors are neglected.
2. The viscous and Coulomb frictions of motors and differentials in the transmission are all neglected.

Therefore, when the vehicle accelerates:

$$\text{Front wheels}: \quad T_{mf}n_g = \frac{2I_w\dot{\omega}_f}{n_g} + 2F_{xf}r_t \tag{12}$$

$$\text{Rear wheels}: T_{mi} = I_w\dot{\omega}_i + F_{xr}r_t, \quad i = r, l \tag{13}$$

where n_g is the reduction ratio of gearbox; I_w is the wheel inertia; ω_f is the speed of the front traction motor; ω_r and ω_l represent, respectively, the speed of the right and left rear in-wheel motors. In the steady state for a small steer angle and by neglecting frictions on wheel motors, the yaw moment Equation (9) can be simplified as

$$T_{mr} = T_{ml} + 2r_t M_z / L_w \tag{14}$$

where L_w is the distance between two rear wheels. This equation will be used for determining the real-time torque distribution in the PSO process when the vehicle is cornering. For a straight-line driving, $M_z = 0$ and $T_{mr} = T_{ml}$.

2.5. Battery State of Charge Model

The SOC of each battery pack on the EV is estimated by the following equation:

$$SOC = SOC_i - \frac{\int_0^t I_b dt}{Q_b} \tag{15}$$

where SOC_i is the initial state of SOC, I_b is the battery current, and Q_b is the battery maximum capacity. By a simple internal resistance model, the output power of battery is estimated by:

$$P_b = V_{oc}I_b - I_b^2 R_b \tag{16}$$

where R_b is the internal resistance of the battery and was set at 0.17 Ω for all the battery packs in the MIL simulations in Section 4. The open circuit voltage (OCV) V_{oc} is a function of SOC and is obtained from experiments for each battery pack. Figure 3 illustrates the OCV curve of the front battery pack. Accordingly, the battery current I_b is easily calculated from (16).

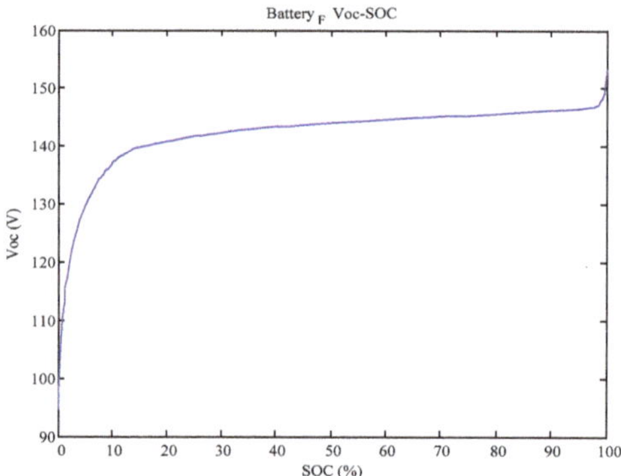

Figure 3. The open circuit voltage and state of charge curve of the front battery pack.

3. Torque and Battery Distribution Strategy

3.1. Torque Distribution Strategy: Particle Swarm Optimization

The total torque, T_t, used to accelerate the vehicle was

$$T_t = (F_x/r_t) = T_{mf}n_g + T_{mr} + T_{ml} \tag{17}$$

From (14), the corresponding yaw moment, M_z, in the steady state of a small steer angle was provided by the differential torque of the left and right in-wheel motors. The maximum and minimum torque ranges of the three traction motors were first determined on the basis of the TNE maps of the three traction motors, wheel angular speeds, SOC of the batteries, and current limits. PSO was then used to determine the best torque distribution for the three traction motors under the constraints of their operation ranges and Equations (14) and (17).

The PSO theory originated by observing the hunting behavior of a swarm of birds or fish. In the process of torque distribution, a particle is a point on a search space of TNE map. As shown in Figure 2a,b, three swarms of particles were initially distributed at random on the three TNE maps of the traction motors, and each swarm had N particles. Each particle had a position state, which was defined as the torque where the particle was located. The pedal command given by the EV driver was their common target at a specific vehicle speed.

Each particle at its initial position in the search space determined its best direction, and all the particles approached their common target with the minimal global effort.

$$\eta = \frac{T_{mf}\omega_{mf}}{\eta_{mf}\left(T_{mf}, \omega_{mf}\right)} + \frac{T_{ml}\omega_{ml}}{\eta_{ml}\left(T_{ml}, \omega_{ml}\right)} + \frac{T_{mr}\omega_{mr}}{\eta_{mr}\left(T_{mr}, \omega_{mr}\right)} \tag{18}$$

During the PSO process, the N particles were renewed through J generations and reached the target of minimal energy consumption in the end. After the least energy consumption converged in each generation, the particles were updated according to their own best and swarm best solutions as

$$\Delta T_{mf,i}^{j+1} = wT_{mf,i}^{j+1} + rand_1 \times c_{L1}\left(P_i^j - T_{mf,i}^j\right) + rand_2 \times c_{L2}\left(G^j - T_{mf,i}^j\right) \tag{19}$$

$$T_{mf,i}^{j+1} = T_{mf,i}^j + \Delta T_{mf,i}^{j+1} \tag{20}$$

where the sub-index i stands for the ith particle; j stands for the generation number; c_{L1} and c_{L2} are learning factors; G represents the swarm's best known solution of all the particles; P_i represents its own best known solution of the ith particle; $rand_1$ and $rand_2$ are random values between 0 and 1; and w is the inertia weight. At the final generation, the output torques of the rear left and right motors were calculated:

$$T_{ml} = \frac{T_t - T_{mf}\,n_g}{2} - \frac{r_t M_z}{L_w} \tag{21}$$

$$T_{mr} = \frac{T_t - T_{mf}\,n_g}{2} + \frac{r_t M_z}{L_w} \tag{22}$$

Details of the original work of the real-time torque-distribution strategy by PSO for a pure EV with three traction motors were described in [16].

3.2. Torque Distribution Strategy: Priority Torque Ratio in Front and Rear Motors

For comparison, a priority torque ratio (PTR), P_r, can be assigned from 0 to 1 to the front motor with respect to the rear motors. For example, $P_r = 1$ means that the front motor takes the first priority for delivering the torque within its feasible range for driving the EV, while $P_r = 0$ expresses that the rear motors have the top priority for delivering the torque over the front motor. Therefore, the torque given by the front motor is calculated as

$$T_{mf} = \left(T_{mf,max} - T_{mf,min}\right)P_r + T_{mf,min} \tag{23}$$

where $T_{mf,max}$ and $T_{mf,min}$ are, respectively, the upper and lower limits of the front motor at a certain speed. If T_{mf} is calculated higher than or equal to the total torque T_t required for acceleration, the front motor will provide T_t as demanded. If T_{mf} is calculated lower than T_t even though $P_r = 1$, the rear motors must be responsible for delivering the rest portion of torque according to (21) and (22).

3.3. Battery Energy Consumption

The energy consumption and balance of the three battery packs were investigated when the torque distribution was executed by the PSO or PTR strategy. In this study, the urban driving cycle (UDC) of the New European driving cycle (NEDC) was used because of the limited motor speeds and battery voltage. Table 2 presents two simulation results for the energy consumption for the three battery packs after the EV drove (1) on a straight road and (2) clockwise on a circular path with a 100 m radius using the proposed PSO and PTR torque distribution strategies. For the cases of $P_r = 1$ and 0.75, the front traction motor had the highest priority by delivering power over the rear motors, and the rear battery restored more power from regenerative braking than that consumed for driving. This was an example of negative battery energy consumption.

Table 2. Energy consumption of battery packs.

Strategy		Battery Energy Consumption [Wh]			
		Front	Rear Right	Rear Left	Total
UDC on a Straight Road					
	PSO	223.2	37.76	37.76	298.72
	$P_r = 0$	44.00	158.4	158.4	361.16
	$P_r = 0.25$	55.20	150.8	150.4	356.68
PTR	$P_r = 0.5$	91.60	123.2	123.2	338.44
	$P_r = 0.75$	338.8	−11.88	−11.88	315.12
	$P_r = 1$	351.2	−19.72	−19.72	311.92
UDC Clockwise along a Circular Path (Radius 100 m)					
	PSO	285.6	96.00	35.36	417.08
	$P_r = 0$	93.20	216.8	158.0	467.92
	$P_r = 0.25$	113.2	205.2	146.0	464.28
PTR	$P_r = 0.5$	179.2	166.4	107.6	453.44
	$P_r = 0.75$	429.2	32.32	−14.28	447.36
	$P_r = 1$	447.2	22.20	−16.08	453.40

For the three battery packs, energy consumption was found to be unbalanced, thereby causing an unbalanced SOC. If any of the three battery packs was depleted, the torque distribution strategy would fail. This could cause a serious deterioration in vehicle maneuverability and stability.

3.4. Torque and Battery Distribution (TBD) Strategy

Figure 4 shows the flowchart for the torque and battery distribution (TBD) strategy. Before the EV started from rest, the SOC of the front, rear right, and rear left battery packs was respectively measured as SOC_F, SOC_L, and SOC_R. The SOC gap and SOC ratio were calculated:

1. SOC gap (SOC_g): The SOC gap was defined as the difference between the SOC of the front battery pack and the lower SOC of the two rear battery packs. It was negative when the front battery had less power remaining than the rear batteries, and it was positive when the front battery had more

power than the rear batteries. In applications, a default value $SOC_g < -\kappa\%$ ($0 < \kappa < 2$) can be assigned to determine the torque distribution mode (Figure 4).

2. SOC ratio (SOC_r): The SOC ratio was defined as the ratio of energy consumption in terms of the SOC between the front battery pack and the two rear battery packs. On the basis of the simulation of straight road driving under the PSO strategy, as indicated in Table 2, the SOC_r converged to 2.94 for the long-term operation of the UDCs. In applications, a default value, ρ (< 2.94), was assigned to determine the torque distribution mode under the TBD strategy.

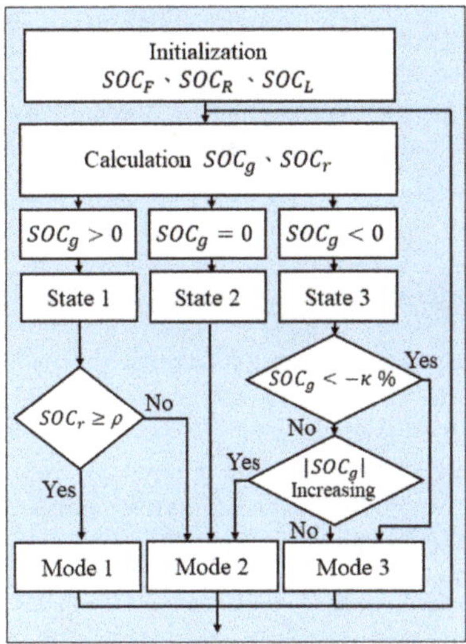

Figure 4. Flowchart of the torque and battery distribution strategy.

Through the use of SOC_g, the state of balance (SOB) of the three battery packs was investigated. The three states of battery balance are described:

1. State 1: The SOC of the front battery pack was higher than that of any of the rear battery packs.
2. State 2: The SOC of the front battery pack was equal to that of any of the two rear battery packs.
3. State 3: The SOC of the front battery pack was lower than that of the rear battery pack.

Three torque distribution modes were then determined by the SOC gap (SOC_g):

1. Mode 1: $P_r = 1$ was proposed when the front motor had the first priority for delivering the torque under the condition of $SOC_g > 0$ and $SOC_r \geq \rho$.

At Mode 1, the SOC of the front battery pack was much higher than that of the rear battery packs. When the SOC_r was larger than 2.94, the SOC_g could increase continuously, and the battery balance could worsen even though the PSO strategy was executed. It was better for the front battery pack to reach a balance between the front and rear batteries.

2. Mode 2: The PSO strategy was prescribed when there was not much difference in the SOC of the three battery packs under the following conditions: $SOC_g = 0$ or ($SOC_g > 0$ and $SOC_r < \rho$) or ($0 > SOC_g \geq -\kappa\%$ and $|SOC_g|$ was increasing).

Because the PSO strategy is superior to the PTR strategy for distributing the torque among the traction motors, it should be used whenever the difference in the SOC of the battery packs is negligible. For example, $SOC_g = 0$ is an ideal case in which all of the batteries are in balance, and the PSO strategy can save more energy than the other PTR strategies when distributing the torque. With $SOC_g > 0$ and $SOC_r < \rho$, the amount of power stored by the front and rear batteries was similar and sufficient. It was also an appropriate situation for torque distribution under the PSO strategy.

When $0 > SOC_g \geq -\kappa\%$, there was not much difference in the energy storage of the battery packs, and it was still safe to execute PSO even though the SOC gap was increasing.

3. Mode 3: $P_r = 0$ was proposed when the rear motors took top priority for delivering more the torque than the front motor under the following conditions: $SOC_g < -\kappa\%$ or $(0 > SOC_g > -\kappa\%$ and $|SOC_g|$ was decreasing).

At Mode 3, the SOC of the rear battery packs was much higher than that of the front battery. The consumption in the rear batteries had top priority so that the balance in the batteries could be restored. After the SOC_g was restored within $[0, -\kappa\%]$, Mode 3 remained in operation because the SOC gap continued to decrease until the best balance state 2 was achieved. This avoided frequent shifts between Modes 2 and 3.

4. Experiments

4.1. Model-in-the-Loop Simulations

In Figure 5, the TBD strategy was simulated on a model-in-the-loop (MIL) platform. MATLAB Simulink was applied to model the TBD strategy, battery model, driver model, slip ratio control (SRC), and direct yaw moment control (DYC), while CarSim provided the vehicle dynamics. The driver model simulated human driver behavior by a proportional-integral (PI) controller. The SRC was responsible for stabilizing vehicle motion through the tractive control system (TCS) during acceleration and the anti-lock brake system (ABS) during deceleration. The TBD strategy was performed after vehicle safety was confirmed. Because of the limited motor speeds and battery voltages, only the UDC part of NEDC was used for MIL simulations.

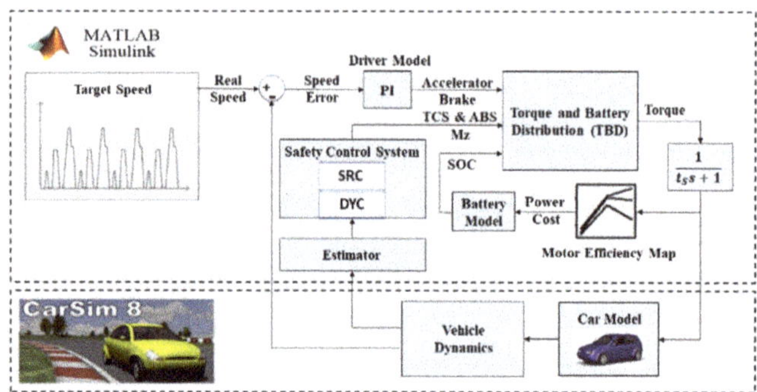

Figure 5. Simulation block diagram of torque and battery distribution strategy.

Both the straight and circular road tests were simulated. Figure 6a shows the SOC records of the three battery packs during the TBD process for the EV driving a clockwise cornering on a circular path of radius 100 m for 3.5 UDCs of the NEDC. The indices ρ and κ were assigned at 1.0056 and 1, respectively.

At the beginning, the SOC of the front battery (91%) was higher than that of both rear batteries (90%). In addition, $SOC_r > 1.0056$, the torque distribution of Mode 1, was executed until the SOC_r reduced to 1.0056 at approximately 170 s, where the PSO of Mode 2 was executed for torque distribution.

The torque distribution mode shifted from Mode 2 to Mode 3 when the SOC_g was less than -1% at approximately 1375 s. The torque distribution mode shifted back to Mode 2 about 1575 s when the SOC gap was reduced. The SOC of the front and rear batteries remained within a SOB by shifting the torque distribution mode during the driving cycle. The torque distribution histories from three traction motors are presented in Figure 6b.

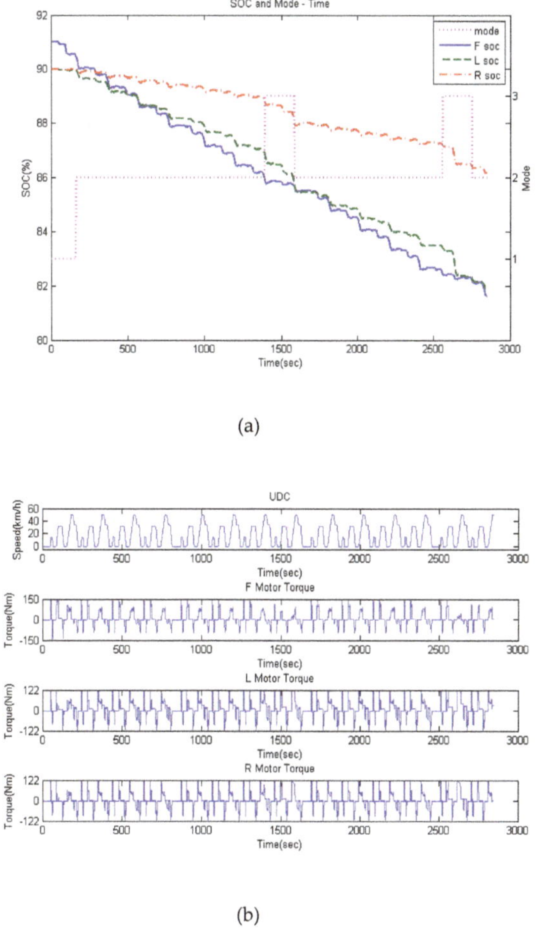

(a)

(b)

Figure 6. (a) The state of charge records for the three battery packs and (b) torque distribution from the front, rear left, and rear right traction motors during the torque and battery distribution process for the electric vehicle driving on a circular path of radius 100 m and using the urban driving cycle of the New European Driving Cycle.

It was also interesting to compare the differences of the energy economy of the proposed TBD strategy by having the battery energy storage in balance and using other torque distribution strategies without balancing the batteries. In these simulations, the initial SOC of each of the three battery packs was 90%, and their corresponding amounts of energy are 10.45, 5.2, and 5.2 kWh for the front, rear right, and rear left battery packs, respectively. The EV stopped when any of the batteries was depleted. It was found that both the rear batteries were exhausted soon for $P_r = 0$, 0.25, and 0.5 when the rear motors took higher priority for delivering more torque than the front motor; while the front battery was depleted soon for $P_r = 0.75$ and 1 when the front motor had the first priority for delivering the torque to the EV.

Energy consumption efficiency was defined as the ratio between the total energy consumption and the initial battery energy. The energy consumption rate was defined as the total energy consumption per travel distance.

Table 3 shows the energy consumption results for both the circular and straight road simulations. For clockwise cornering on a circular path of radius 100 m during the UDCs, the proposed TBD strategy had a travel distance: 142.6 km, which was 7.7% higher than the PSO strategy without the balancing of the SOC of the battery packs. The torque distribution strategy under PSO without battery balancing had a better energy consumption rate at 104.5 Wh/km than the TBD strategy, but the front battery was delpeted after 132.4 km, and the energy consumption efficiency was 73.6%. The rear battery packs had 26.4% energy remaining when the EV stopped.

Table 3. Torque and battery distribution strategy results.

Strategy	Energy Consumption (kWh)	Travel Distance (km)	Energy Consumption Rate (Wh/km)	Energy Consumption Efficiency (%)
Clockwise UDC along a Circular Path (Radius 100 m)				
TBD	16.52	142.6	115.9	88.0
PSO	13.89	132.4	104.9	73.5
$P_r = 0$	10.11	85.9	117.7	53.3
$P_r = 0.25$	11.10	90.6	122.5	58.6
$P_r = 0.5$	12.76	111.9	114.1	67.4
$P_r = 0.75$	9.91	88.2	112.4	52.2
$P_r = 1$	9.64	84.5	114.0	50.8
UDC on a Straight Road				
TBD	18.75	214.4	87.4	99.9
PSO	12.59	167.6	75.1	67.1
$P_r = 0$	10.66	117.4	90.8	56.8
$P_r = 0.25$	11.08	123.5	89.7	59.0
$P_r = 0.5$	12.85	151.0	85.1	68.5
$P_r = 0.75$	8.74	110.3	79.3	46.6
$P_r = 1$	8.35	106.4	78.4	44.5

On the straight road, the torque distribution strategy under PSO without battery balancing exhibited the best energy consumption rate: 75.21 Wh/km. However, the front battery was depleted after 167.6 km, and energy consumption efficiency was 67.5%. Thus, the rear battery packs had only 32.5% energy remaining when the EV stopped. The proposed TBD strategy of partly using torque distribution by PSO had the highest driving range: 214.4 km, i.e., approximately 248 UDCs. This was attributed to the SOC of the three battery packs being kept in balance to avoid unexpected battery depletion. Therefore, approximately 28% more driving range was extended by the TBD strategy than by the PSO strategy without battery balancing.

It was also found in the simulation that the battery energy was fully utilized by the TBD strategy for the EV on a straight road of UDC. The 99.9% energy consumption efficiency for the TBD strategy was calculated in Table 4. The energy of three battery packs can be effectively distributed and utilized to extend the driving range, when the SOC gap of the three battery packs remains within a prescribed limit during vehicle operation.

Table 4. Energy consumption efficiency for the EV on a straight road of UDC by the TBD strategy.

Battery	Front	Rear Left	Rear Right
Energy capacity (kWh) @ 100% SOC	10.45	5.2	5.2
Initial SOC (%)	90	90	90
Final SOC (%)	0	1.75	1.75
Initial battery energy (kWh)	$A = (10.45)(0.9) + (5.2)(0.9)(2) = 18.77$		
Total energy consumption (kWh)	$B = (10.45)(0.9-0) + (5.2)(0.9-0.0175)(2) = 18.75$		
Energy consumption efficiency	$A/B = 18.75/18.77 = 99.89\%$		

Other strategies with a PTR, P_r from 0 to 1, presented lower travel distances and less energy consumption efficiency than observed for the proposed TBD strategy.

4.2. Hardware-in-the-Loop Simulations

Figure 7 presents the architecture of hardware-in-the-loop experiment. A Mitsubishi Colt-Plus was retrofitted with a 15-kW radial-flux PMSM and a 144-V battery pack for front wheels and two 7-kW axial-flux PMSMs and two 72-V battery packs for rear wheels. This EV was set up on a Horiba MAHA-AIP ECDM-48 emission chassis dynamometer. The maximum test speed was 200 km/h, the maximum power absorbing was 150 kW at 100 km/h, the maximum tractive force was 5400 N for light duty and 6750 N for heavy duty, and the maximum vehicle inertia simulation was 4540 kg for 4-wheel drives.

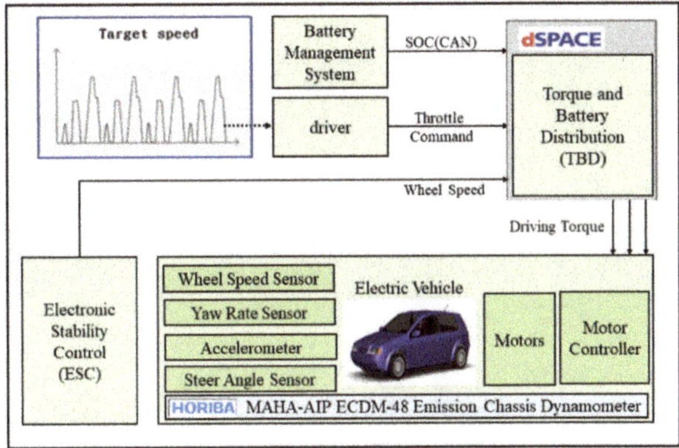

Figure 7. Architecture of the hardware-in-the-loop experiment.

In experiments, the battery SOC information and throttle command were received by a real-time rapid control prototyping unit dSPACE MicroAutoBox II, in which the TBD strategy was built to determine the torque command for each motor. This MicroAutoBox took a role of vehicle control unit with an 800-MHz processor, 18-MB main memory, 16-MB flash memory, and dual CAN interfaces.

In the HIL experiment, the vehicle followed the complete NEDC but the maximum speed at EUDC was restricted at 40 km/h in the HIL experiment. The initial SOCs of the front, rear left, and rear right batteries were 94.8%, 93.6%, and 95.5%, respectively. In order to make the experiment efficient, driving modes were shifted at $SOC_g = -\kappa\% = -1\%$ and $SOC_r = \rho = 1.006$, according to the flowchart of the TBD strategy in Figure 4.

Figure 8 illustrates the time history of mode, state, SOC_r, and SOC_g and the corresponding torque distribution histories of the front, rear left, and rear right motors during the TBD process in the hardware-in-the-loop experiment. During the first 95 s, the SOC of the front battery pack was higher than the SOC of anyone of the rear battery packs. Therefore, State 1 ($SOC_g > 0$) was identified and SOC_r was larger than 1.006, the front motor took the first priority of delivering torque and Mode 1 ($P_r = 1$) was executed. Between 95 and 200 s, it was still at State 1 ($SOC_g > 0$) but SOC_r was less than 1.006, Mode 2 was executed with the PSO strategy.

Between 200 and 300 s, State 2 ($SOC_g = 0$) was identified, i.e., the SOC of the front battery pack was equal to any one of the two rear battery packs. Mode 2 with the PSO strategy was working. Between 300 and 485 s, the SOC of the front battery pack was lower than the SOC of the rear battery packs, and State 3 ($SOC_g < 0$) was identified. Because $0 > SOC_g \geq -1\%$ and $|SOC_g|$ was increasing, Mode 2 remained.

When SOC_g was less than -1% at State 3 between 485 and 585 s, the rear battery packs and motors started to take their top priority for delivering more power than the front battery and motor, and Mode 3 ($P_r = 0$) was executed. Between 585 and 680 s, the SOC gap was between -1% and 0%, but $|SOC_g|$ was decreasing, Mode 3 remained.

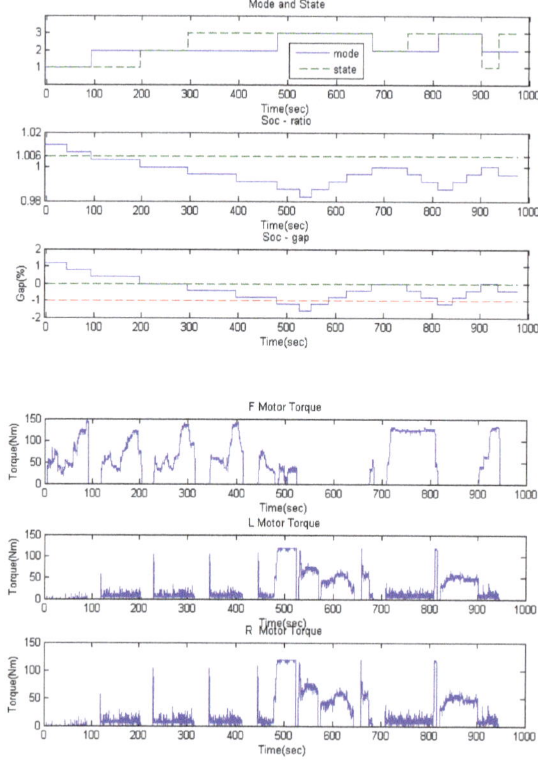

Figure 8. Histories of the mode (solid line), state (dashed line), state of charge (SOC) ratio, SOC gap and the torque distributions of the front, rear left, and rear right motors during torque and battery distribution process for the electric vehicle in the hardware-in-the-loop experiment.

The corresponding SOCs of the three battery packs is shown in Figure 9. In the first 500 s, the front battery provided all power to the vehicle. After 500 s, the rear and front batteries powered the vehicle alternatively, so that the SOC gap would always remain in 1% as expected.

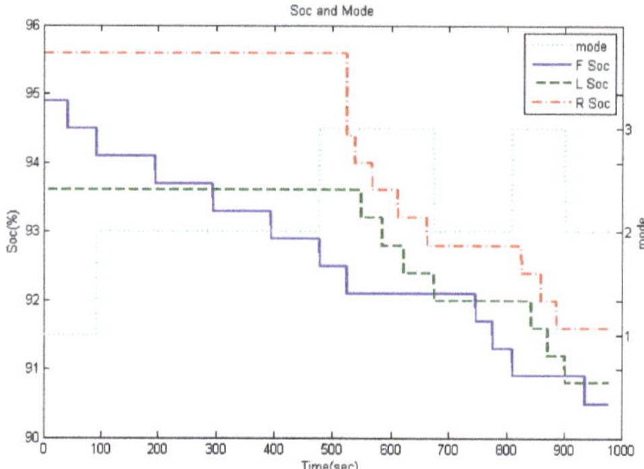

Figure 9. Histories of mode and the states of charge of the front, rear left, and rear right battery packs during torque and battery distribution process for the electric vehicle in the hardware-in-the-loop experiment.

4.3. Road Tests

The road test was executed on the Industrial Technology Research Institute (ITRI) campus. The total driving distance was about 5.4 km, during which the road slope varied and the maximum speed was 30 km/h. The driving curve is shown in Figure 10. The initial SOCs of the front, rear left, and rear right batteries were 72.8%, 70%, and 71.2%, respectively. Driving modes were shifted at $SOC_g = -\kappa\% = -1\%$ and $SOC_r = \rho = 1.02$, according to the flowchart of the TBD strategy in Figure 4. Figure 11 illustrates the time history of mode, state, SOC_r, SOC_g, and the corresponding torque distributions.

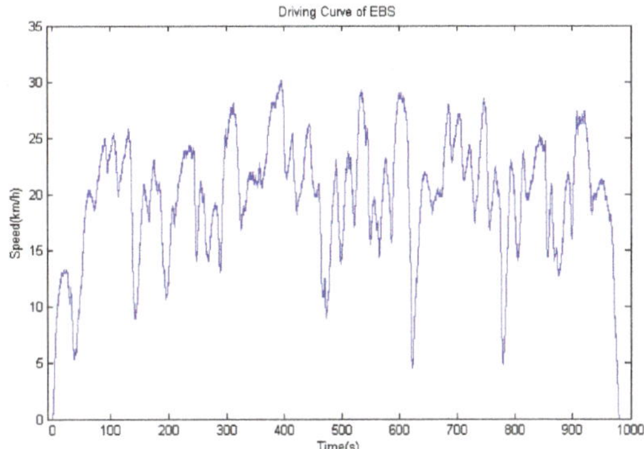

Figure 10. Driving speed curve for the torque and battery distribution strategy in the road test on the campus of Industrial Technology Research Institute.

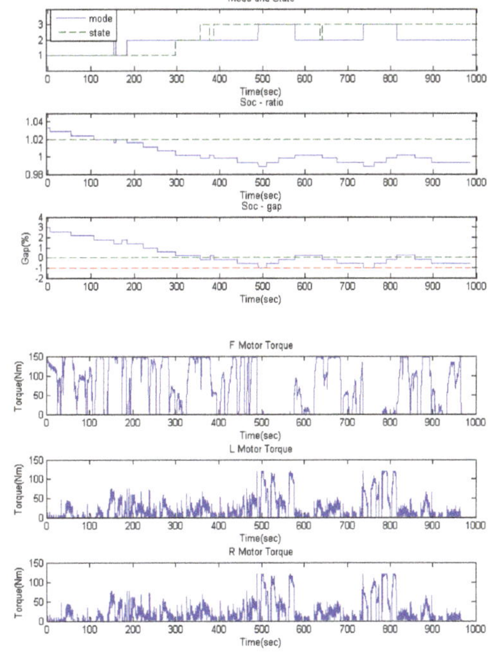

Figure 11. Histories of the mode (solid line), state (dashed line), state of charge (SOC) ratio, SOC gap and the torque distributions of the front, rear left, and rear right motors during torque and battery distribution process for the electric vehicle in the road test.

Similar to the result of HIL test, the front battery pack took the first priority to supply power to drive the EV at Mode 1 ($P_r = 1$) during the first 180 s. Because the EV moved upslope approximately at 150 s, two rear motors delivered extra torque. Between 180 and 480 s, Mode 2 with the PSO strategy worked when the SOC of the front battery was lower than that of the rear battery packs. The front motor still provided the major torque until the SOC gap was less than −1% ($SOC_g < -1\%$) while Mode 3 ($P_r = 0$) was executed. Then, Modes 2 and 3 shifted alternatively. The SOC difference of the three battery packs was finally kept within 1%, as shown in Figure 12.

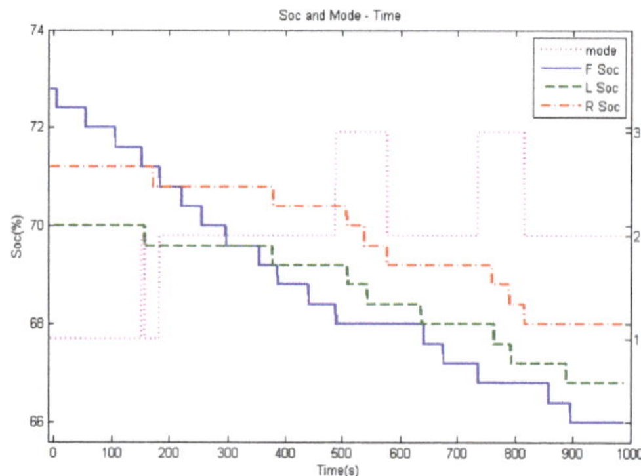

Figure 12. The variation of SOC of the three battery packs in the road test.

The two strategies with a PTR, $P_r = 0$ and $P_r = 1$, were also executed in the road test. The total energy consumption and the consumption of each battery pack are given in Table 5. It shows that the energy consumption rate of the TBD strategy was lower than those of $P_r = 0$ and $P_r = 1$. It means that with the same battery energy, the EV has a longer driving range by the proposed TBD strategy than that by other PTR strategies. For example, the TBD strategy extended 11% more driving range than the PTR strategy when $P_r = 0$, and the TBD strategy extended 23.5% more driving range than the PTR strategy when $P_r = 1$.

Table 5. Battery energy consumption in road tests.

Strategy	Battery Energy Consumption [Wh]			Driving Range (m)	Energy Consumption Rate [Wh/km]
	F	**RL**	**RR**		
TBD	797	149	149	5464	200
$P_r = 0$	179	515	510	5424	222
$P_r = 1$	1318	14	18	5470	247

F: Front battery, RL: Rear left battery, RR: Rear right battery.

5. Conclusions

A novel TBD strategy has been proposed for the EV with three independent traction motors and battery packs. Upon acceleration of the EV, the demanded torque was provided by all three traction motors together at their highest efficiency under the PSO strategy for saving battery energy. Simultaneously, the SOC of the three battery packs had to be kept in balance to avoid any unexpected battery depletion and to improve the EV's driving range. Thus, a combination of PSO and the PTR for torque distribution strategy was applied to compromise between energy saving and energy balance. On the basis of the model-in-the-loop simulations, the proposed TBD strategy shows better travel distance and the higher energy consumption efficiency than a pure PSO method or PTR strategies. Similar results were also proved on a real vehicle for hardware-in-the-loop experiments on dynamometer and road tests. The road test proved that the TBD strategy extended 11% and 23.5%

more driving range than other PTR strategies, when $P_r = 0$ and $P_r = 1$, respectively. The proposed TBD strategy is promising for extending the driving range of an EV with multiple traction motors and battery packs with an improved energy consumption efficiency.

Author Contributions: Y.-H.T. and Y.-P.Y. conceived and design the experiments; Y.-H.T. performed the experiments and analyzed the data; Y.-P.Y. wrote the paper. All authors have read and agreed to the published version of the manuscript.

Funding: This research was funded by the Ministry of Science and Technology of Taiwan, Republic of China under contract MOST 106-2221-E-002-064.

Acknowledgments: The authors acknowledge the financial support of the Ministry of Science and Technology of Taiwan, Republic of China under contract MOST 106-2221-E-002-064.

Conflicts of Interest: The authors declare no conflicts of interest.

References

1. Li, W.; Fu, Y.; Liu, T.; Chu, P.; Wang, J.; Chen, H. Battery equalization based on state of charge. In Proceedings of the 2nd International Conference on Systems and Informatics, Shanghai, China, 15–17 November 2014; pp. 159–163.
2. Gallardo-Lozano, J.; Romero-Cadaval, E.; Jalakas, T.; Hõimojaet, H. A battery cell balancing method with linear mode bypass current control. In Proceedings of the 14th Biennial Baltic Electronic Conference, Tallinn, Estonia, 6–8 October 2014; pp. 245–248.
3. Huang, W.; Abu Qahouq, J.A. Energy sharing control scheme for state-of-charge balancing of distributed battery energy storage system. *IEEE Trans. Ind. Electron.* **2015**, *62*, 2764–2776. [CrossRef]
4. Pham, V.L.; Tung, N.T.; Tran, D.H.; Vu, B.; Choi, W. A new cell-to-cell fast balancing circuit for lithium-ion batteries in electric vehicles and energy storage system. In Proceedings of the IEEE 8th International Power Electronics and Motion Control Conference, Hefei, China, 22–26 May 2016; pp. 2461–2465.
5. Dung, L.R.; Lin, C.T.; Yuan, H.F. An anti-racing PWM-based battery pack equalization. In Proceedings of the 14th International Conference on Control, Automation and Systems, Seoul, Korea, 22–25 October 2014; pp. 1244–1248.
6. Jin, F.; Wang, M.; Hu, C. A fuzzy logic based power management strategy for hybrid energy storage system in hybrid electric vehicles considering battery degradation. In Proceedings of the IEEE Transportation Electrification Conference and Expo, Dearborn, MI, USA, 26–29 June 2016; pp. 1–7.
7. Akar, F.; Tavlasoglu, Y.; Vural, B. An energy management strategy for a concept battery/ultracapacitor electric vehicle with improved battery life. *IEEE Trans. Transport. Electrific.* **2017**, *3*, 191–200. [CrossRef]
8. Menon, R.; Azeez, N.A.; Kadam, A.H.; Williamson, S.S. Carrier based power balancing in three-level open-end drive for electric vehicles. In Proceedings of the IEEE 12th International Conference on Compatibility, Power Electronics and Power Engineering, Doha, Qatar, 10–12 April 2018; pp. 1–6.
9. Tanaka, Y.; Tsuruta, Y.; Nozaki, T.; Kawamura, A. Proposal of ultra high efficient energy conversion system (HEECS) for electric vehicle power train. In Proceedings of the IEEE International Conference on Mechatronics, Nagoya, Japan, 6–8 March 2015; pp. 703–708.
10. Rossi, C.; Pontara, D.; Bertoldi, M.; Casadei, D. Two-motor, two-axle traction system for full electric vehicle. *World Electr. Veh. J.* **2016**, *8*, 25. [CrossRef]
11. Yin, D.; Shan, D.; Chen, B.C. A torque distribution approach to electronic stability control for in-wheel motor electric vehicles. In Proceedings of the International Conference on Applied System Innovation, Okinawa, Japan, 28 May–1 June 2016; pp. 1–4.
12. Zhai, L.; Sun, T.; Wang, J. Electronic stability control based on motor driving and braking torque distribution for a four in-wheel motor drive electric vehicle. *IEEE Trans. Veh. Technol.* **2016**, *65*, 4726–4739. [CrossRef]
13. Dizqah, A.M.; Lenzo, B.; Sorniotti, A.; Gruber, P.; Fallah, S.; Smet, D.J. A fast and parametric torque distribution strategy for four-wheel-drive energy-efficient electric vehicles. *IEEE Trans. Ind. Electron.* **2016**, *63*, 4367–4376. [CrossRef]
14. Fujimoto, H.; Harada, S. Model based range extension control system for electric vehicles with front and rear driving braking force distributions. *IEEE Trans. Ind. Electron.* **2015**, *62*, 3245–3254. [CrossRef]

15. Sun, H.; Wang, H.; Zhao, X. Line braking torque allocation scheme for minimal braking loss of four-wheel-drive electric vehicles. *IEEE Trans. Veh. Technol.* **2019**, *68*, 180–192. [CrossRef]
16. Yang, Y.P.; Shih, Y.C.; Chen, J.M. Real-time torque-distribution strategy for a pure electric vehicle with multiple traction motors by particle swarm optimization. *IET Elect. Syst. Transp.* **2015**, *6*, 76–87. [CrossRef]
17. Yang, Y.P.; Chen, W.C. Coupled energy-saving and safe-driving strategy for an electric vehicle driven by multiple motors. In Proceedings of the 30st International Electric Vehicle Symposium & Exhibition, Stuttgart, Germany, 9–11 October 2017; pp. 1–10.
18. Tseng, Y.H.; Yang, Y.P. An energy saving strategy of torque and battery distribution for an electric vehicle driven by multiple traction motors. In Proceedings of the Electric Vehicles International Conference & Show, Bucharest, Romania, 3–4 October 2019.

Article

Development of a Comprehensive Model for the Coulombic Efficiency and Capacity Fade of LiFePO$_4$ Batteries under Different Aging Conditions

Ting-Jung Kuo

Department of Engineering Science and Ocean Engineering, National Taiwan University, No. 1, Sec. 4, Roosevelt Rd., Taipei 106, Taiwan; D99525009@ntu.edu.tw; Tel.: +886-02-33665737

Received: 28 September 2019; Accepted: 23 October 2019; Published: 28 October 2019

Abstract: In this paper, a comprehensive model for LiFePO$_4$ batteries is proposed to ensure high efficiency and safe operation. The proposed model has a direct correlation between its parameters and the electrochemical principles to estimate the state of charge (SoC) and the remaining capacity of the LiFePO$_4$ battery. This model was based on a modified Thévenin circuit, Butler–Volmer kinetics, the Arrhenius equation, Peukert's law, and a back propagation neural network (BPNN), which can be divided into two parts. The first part can be represented by the dual exponential terms, responsive to the Coulomb efficiency; the second part can be described by the BPNN, estimating the remaining capacity. The model successfully estimates the SoC of the batteries that were tested with an error of 1.55%. The results suggest that the model is able to accurately estimate the SoC and the remaining capacity in various environments (discharging C rates and temperatures).

Keywords: LiFePO$_4$ batteries; state of charge (SoC); Butler–Volmer equation; Arrhenius; Peukert; coulomb efficiency; back propagation neural network (BPNN)

---

1. Introduction

LiFePO$_4$ is one of the most popular cathode materials for lithium batteries with developmental potential. In recent years, it has become a research topic receiving attention from both academia and industry. The battery possesses advantages such as high coulombic efficiency [1,2], high theoretical capacity (170 mAh·g^{-1}), low cost, and a long cycle life. It is widely employed in many energy storage commercial applications such as electric vehicles and hybrid vehicles [1–4]. However, under variable working conditions, such as different temperatures, depths of discharge, charging and discharging rates, and numbers of cycles, the aging of the battery would speed up, leading to error in estimating the battery strength. Much research has been conducted to investigate the mechanisms of battery aging [5–9]. However, it is still a major challenge to quantify these factors' influence on battery performance.

Since battery aging is a process of a set of complex mutual interactions, aging will be sped up by a high charging/discharging rate, a working environment of high or low temperature and over-discharge. When the environmental temperature is higher than 45 °C, the efficiency of the anode will be influenced by surface formation and structural degeneration, which becomes a significant factor for the aging of LiFePO$_4$ batteries. The battery life at 55 °C is 1/7 of that at 25 °C [10]. Rising temperature clearly promotes the dissociation of the electrolyte and iron ions dissolving [11], leading to the lithium ions being rapidly used up and the degradation of the anode's structure, accelerating the aging of the LiFePO$_4$ battery. In contrast, when the environmental temperature is below 0 °C, the rate of electrochemical reactions drops. This includes ion conduction in the electrolyte, the transmission of lithium ions through solid state electrolyte interface, charge transferral at interface, and the rate of spread of the solid-state lithium. As a result, there are losses in energy and strength [12–14]. The main phenomenon causing this kind of rapid capacity fading is known as lithium electroplating, occurring when a battery is always under harsh operating conditions, such as low temperature, strenuous charge, or overcharging [15–17]. Among the various degradation mechanisms occurring in the Li-ion batteries, lithium plating is considered as one of the most detrimental. Because the lithium plating not only catalyzes further degradation, but also affects the safety of the battery operation [18,19]. The discharging current also has a large influence on the degradation of battery performance. The work in [20] has shown from simulation results that under high discharge rate, the battery shows a poorer cycle life. Since joule heating and electrochemical reaction heat are released in the battery in charging and discharging conditions, the two exhibit a positive relationship with the current strength.

Electrode tabs are the small metallic strips that are welded onto the current collectors without active materials [21]. When the battery is charged or discharged, the current density near the electrode tabs is higher, which means the temperature of the active materials inside the battery is also higher near the electrode tabs. Uneven temperature distribution will lead to rapid aging of some of the battery's active materials; thus, the available capacity drops. When batteries are sealed into battery packs or are under a high temperature working environment, temperature difference between areas near the electrode tabs and other parts will become even more pronounced. The main reason being a higher temperature will accelerate the rate of rapid secondary reactions of the cathode and the solid electrolyte interface (SEI) growth rate [20]. Conversely, when the batteries are connected in series and sealed into a battery pack, uneven temperature distribution will accelerate aging of active materials locally among the batteries, leading to a drop in capacity [12,20]. This will finally lead to capacity imbalance of battery units in the battery pack, causing malfunction [22].

From the above description, it can be seen that environmental temperature has a large influence on battery aging [23]. The uneven heating produced from a high rate of discharge will also lead to aging of some of the internal materials in the battery, speeding up the aging process [24]. Thus, only when these causes are considered will the batteries operate safely and efficiently.

State of charge (SoC) is defined as the available capacity and expressed as a percentage of the battery's rated capacity. SoC estimation based on the ampere-hour method is calculated using

$$SoC_{(t)} = SoC_{(t_0)} - \frac{1}{C_{N(t)}} \int_{t_0}^{t} (\eta_c \cdot i(t) - S_d) dt, \tag{1}$$

where $SoC_{(t)}$ is the SoC at time t; $SoC_{(t_0)}$ is the initial value; C_N is the rated capacity; η_c is the coulombic efficiency; S_d is the self-discharge rate; and $i(t)$ is the current, which is positive during the discharging process and negative during the charging process.

In Equation (1), coulombic efficiency (CE) can be used to estimate the relationship between SoC and cycle life [25–27]. For LiFePO$_4$ batteries under room temperature, CE > 0.994 and the self-discharge rate is lower than 5% per month [10]. If temperature and self-discharge are not considered, the equation can be simplified by assuming $\eta_c = 1$ and $S_d = 0$. However, temperature change strongly affects CE. When used under low temperature environments and for some high power applications such as electric vehicles, there will be safety issues. In addition, the capacity will fade as the number of cycles increase, causing considerable errors in SoC estimation.

In order to correct the measurement errors of the battery, the work in [28] proposed practical state-of-charge. This definition uses a practical operational capacity, instead of the manufacturer's rated capacity, as the maximum amount of charge. It can be expressed as

$$SoC_N = \frac{Q_{available}}{C_{max,p}}, \tag{2}$$

where $C_{max,p}$, represents the maximum practical capacity as measured from the operating battery at the current time. $C_{max,p}$ may fade over time, due to the effect of battery aging.

During the charging process of lithium batteries, a SEI is formed, leading to loss of active lithium ions and lowering of capacity. State of health (SoH) is another important method of estimation for battery health management. It is the direct indication of the health condition of the battery system. A battery's SoH normally ranges within 0–100%, but when it is new, SoH can be slightly larger than 100% due to product variations [29]. In this paper, to avoid confusion, the SoH_N of the battery refers to the ratio of battery discharge capacity to new battery rated capacity under certain conditions. SoH_N can be expressed as

$$SoH_N = \frac{C_{max,p}}{C_N}.$$

(3)

Battery models describe the charging and discharging behaviors of batteries and information such as calculated capacity, state of health, etc. They can be used for preventing overcharging or over discharging. SoC and SoH are both indispensable methods for battery models. The battery models are categorized into electrochemical models and equivalent circuit models [30]. Electrochemical models use equations of physical phenomena and micro-structure of materials to predict battery performance [31–33]. They are often used to study single specific batteries, analyze the phase changes of battery materials, study the influence from porous structure of battery electrode tabs [34], and optimize the tab porosity rate and tab thickness [35]. Simultaneously, electro-chemical models are also used in improving battery design and manufacturing. However, high computational complexity is the major disadvantage. The models also cannot be used for estimating the performance of battery packs consisting of multiple batteries in electric vehicles. On the other hand, equivalent circuit models can describe battery SoC, SoH, and impedance. Together with actually measured currents and voltages, batteries' non-linear dynamic behaviors can be shown [36,37].

In recent years, machine learning has also been employed for calculations to analyze battery states [38]. Through a great deal of training data, the error between predicted and measured battery performance is minimized. The typical algorithm is an artificial neural network (ANN). In terms of precision, it gives stellar results. Parameters can be calibrated to enable preserving fine performance of the battery in its life cycle [39–42]. However, since machine learning is based on mathematical optimization, battery behaviors and potential physical phenomena cannot be linked together.

This research proposes a comprehensive model to describe the discharging behavior of LiFePO$_4$ batteries. In the model, parameters of the discharging behaviors and electrochemical phenomena are intimately related.

The proposed model is based on a modified Thevenin circuit, Butler–Volmer kinetics, the Arrhenius equation, Peukert's law, and a back-propagation neural network (BPNN). The model estimates the CE and the remaining capacity of the battery in various environments. The simplifications of the electrochemistry equations make it possible to integrate it into equivalent circuit models in order to predict the battery's states during real operation.

The structure of the paper is as follows: Section 2 describes the experimental framework, procedures, and the experimental data. Section 3 introduces the battery models and terminology used throughout this paper. The work including battery modeling and its parameter identification are given in Section 4. A comprehensive battery model is presented to simulate the discharging behavior of the batteries. Curve fitting was employed to estimate the SoC and battery's parameters, while the BPNN estimated the SoH. Section 5 describes that battery capacity tests were conducted to verify the accuracy and robustness of the proposed method in various environments. The SoC estimation results are shown and analyzed. Lastly, the conclusions are drawn in Section 6.

2. Experimental Setup and Procedures

The performance of lithium batteries must be consistent under load and temperature variations. Hence, coulombic efficiency and capacity fade of the battery were analyzed experimentally under various environmental conditions to provide information for electric vehicles and other devices, as well as to ensure operational stability. The method of gathering model parameters was to set up 28 LiFePO$_4$ batteries under 7 temperature settings and 4 discharge rates during charging and discharging

400 cycles. The training and testing data for battery models was constructed from this method. The CE was developed to fit the discharging curve dependent on temperature and discharge current. The irreversible capacity fade of the battery was developed by ANN.

2.1. Experimental Process

In this research, all of the lithium-ion batteries were LYS347094S from Taiwan's LYNO Corporation. These batteries' working voltage and capacity are rated at 3.2 V and 10 Ah. In accordance with the data sheet from the manufacturer, the discharging rates under various temperatures of −10, 0, 10, 20, 30, 40, and 50 °C in a temperature-controlled chamber were tested (DBL45 from Taiwan Dengyng Tec Corporation). These LiFePO$_4$ batteries were discharged under rates of 0.5, 1, 2 and 3 C by a Chroma 17020 battery test station. These batteries were cycled 400 times, within the limits recommended by the manufacturer. Information such as capacity attenuation and number of cycles of the batteriesweare computer recorded. The model was constructed in MATLAB software platform to analyze the recorded discharging curve and perform curve fitting. The experimental setup flowchart is shown in Figure 1.

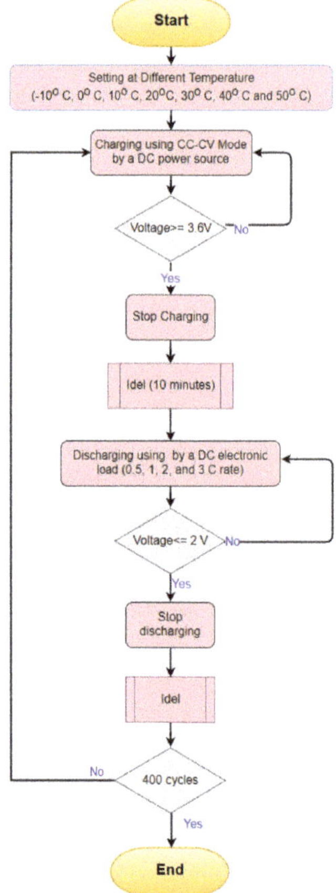

Figure 1. Experimental setup flowchart.

2.2. Coulombic Efficiency Analysis of a Battery under Different Temperatures and Discharge C Rates

Characteristics of the discharging behaviors and voltage variation with time of LiFePO$_4$ battery exhibit a curve with a very flat voltage plateau and a very obvious voltage drop point. When the capacity is ample, the discharging curve is very stable. Once the voltage drop point is reached, the voltage will rapidly drop suggesting that the remaining capacity will be insufficient, as shown from Figure 2a–d. From the figures, the battery voltage decreases drastically, since the curve goes through a knee point, and eventually drops to the cut-off voltage, at which the battery has been exhausted [43–46]. Accord to Faraday's law, the capacity is equal to the integral of current over time. The voltage represents an important characteristic for ample capacity. When the capacity fades, the voltage will be rapid drop, which means that the discharge time before the knee point is significantly reduced [47,48]. The discharge curves between 10 °C and 50 °C can be seen. Before the knee point, the voltage drop is insignificant. Once the knee point is reached, the voltage would rapidly fall [49]. However, the curves of −10 and 0 °C are monotonic. Their knee points are also not obvious. As the discharging rate increases, voltage fluctuation is observable. At the same time, the discharging time of the battery under a low temperature is shorter than that under room temperature and a high temperature. Thus, according to experimental results and comparing LiFePO$_4$ battery discharging between 20 and 50 °C to that of below 10 °C, the latter exhibits worse CE and voltage stability.

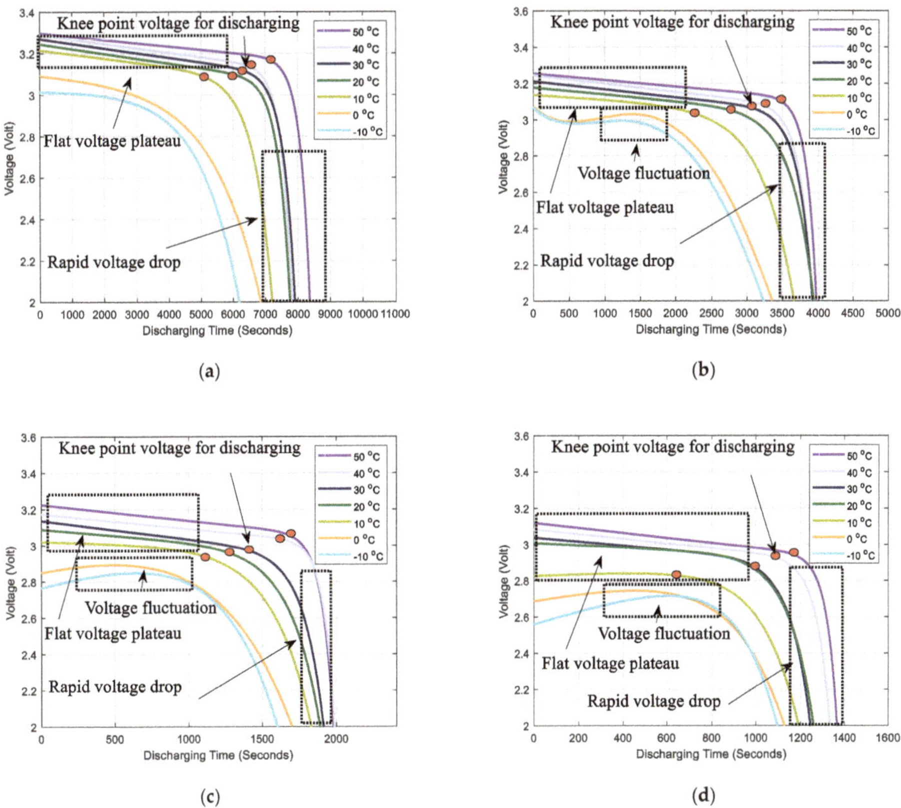

Figure 2. Comparison of the the discharge curves at different discharge C rates under different temperatures for the tested LiFePO$_4$ batteries: (**a**) discharge curves of 0.5 C ate under different temperatures; (**b**) discharge curves of 1 C rate under different temperatures; (**c**) discharge curves of 2 C rate under different temperatures; (**d**) discharge curves of 3 C rate under different temperatures.

2.3. Capacity Fade Analysis of a Battery under Different Temperatures and Discharge Rates for 400 Cycles

Under environmental temperatures of 20 and 30 °C, the results of 400 cycles of different discharge rates are listed in Table 1. The experimental results indicate that at the 400th cycle for rates 0.5, 1, and 2 C, the SoH_N is still higher than 93%. For 3 C rate at the 400th cycle, the SoH_N is lower than 80%. At 20 and 30 °C, the battery performance is very stable and with a very long cycle life. The work in [50] has shown that the discharge capacity measured at 25 °C shows 15.5% loss after 600 cycles; the experimental results are similar to the references [50–52]. Since over 2000 charge and discharge cycles are expected to be undergone by the LiFePO$_4$ battery, considering the experimental cost, the number of battery tests was set to 400 cycles.

Table 1. Capacity change with 400 cycles under different rates of discharge (environmental temperatures 20 and 30 °C).

C Rate	0.5 C		1 C		2 C		3 C	
T (°C)	20	30	20	30	20	30	20	30
Cycle								
1	108.11	108.29	109.00	107.69	106.94	104.28	108.33	101.58
100	105.97	106.15	106.42	106.14	104.06	100.56	103.17	98.08
200	104.61	103.78	102.83	105.00	100.17	97.67	93.75	90.33
300	103.85	100.51	102.11	100.47	96.56	95.22	86.92	85.67
400	102.86	98.97	99.92	98.36	93.22	93.56	78.42	81.83

At 40 °C and 50 °C, the results of 400 cycles of different discharge rates are listed in Table 2. Under high temperature, owing to rapid electrochemical reactions, the CE increases. After 400 cycles at 50 °C and with a 0.5 C rate, SoH_N still has a value of 104.83%. When the discharge rate is raised to 3 C, it can be seen that SoH_N drops rapidly at 400 cycles to only 49.33%. The experimental results indicate that under working conditions of 50 °C and a 3 C rate, the battery life will quickly decay.

Table 2. Capacity change with 400 cycles under different rates of discharge; environmental temperatures 40 and 50 °C.

C Rate	0.5 C		1 C		2 C		3 C	
T (°C)	40	50	40	50	40	50	40	50
Cycle								
1	107.03	114.50	107.17	108.94	108.94	108.22	109.67	105.08
100	105.25	114.31	106.25	108.83	104.61	106.33	106.25	102.58
200	102.12	113.60	104.56	107.44	103.17	99.06	103.50	93.17
300	100.51	110.54	99.33	87.33	97.83	82.00	92.67	77.50
400	99.39	104.83	80.83	69.36	79.50	54.17	78.58	49.33

Under environmental temperatures of −10, 0, and 10 °C, the results of 400 cycles of different discharge rates are listed in Table 3. Under low temperature CE is not high. With −10 °C and 3 C rate as example of the working condition, at the first cycle the SoH_N is only about 91.08%. As the number of cycles increase, the compound influence from low temperature and high discharge rates gradually become obvious. At 400 cycles, the SoH_N is only 26.4%, far lower than the battery life under room temperature. This indicates that under low temperature the CE of LiFePO$_4$ is not high and the capacity fades rapidly.

Table 3. Capacity change with 400 cycles under different rates of discharge; environmental temperatures −10 and 10 °C.

						SoH_N (%)						
C Rate		0.5 C				1 C			2 C			3 C
T (°C)	−10	0	10	−10	0	10	−10	0	10	−10	0	10
Cycle												
1	89.63	95.31	100.0	88.94	95.33	101.5	89.50	93.89	101.2	91.08	92.42	96.67
100	83.63	92.58	97.64	83.64	88.92	99.08	80.11	88.67	97.61	77.83	86.58	93.58
200	81.72	86.04	95.00	81.33	85.03	94.61	75.67	79.83	94.44	67.42	78.67	89.00
300	75.31	81.93	91.85	71.56	73.06	91.83	65.78	70.72	92.39	46.75	70.67	83.83
400	56.17	67.31	89.78	49.17	62.89	88.61	41.67	58.44	87.78	26.42	58.42	79.75

2.4. Parameter Analysis and Comparison

Analyzing the discharge time and voltage plateau curves, it can be observed that at 20, 30, 40, and 50 °C, the influence of high discharge rate to the CE is minor. However, under a high temperature environment, as the Ohmic heating from the increase in number of cycles and high discharge rate rapidly damages the internal materials of the battery; the capacity will quickly decrease. Under low temperature, the rate of electrochemical reactions lowers; by necessity the CE must also fall. However, under an environment of low temperature together with high discharge rate, Ohmic heating will offset some influence from low temperature, allowing the CE to increase slightly. It can be observed from Figure 2a–d that the discharge curves are not stable. However, as the number of cycles increases, due to uneven heating of internal materials of the battery, those close to the battery tip decay more quickly. Hence, there will be a much greater capacity fade compared to prediction.

It can be seen that under the compound influence of environmental temperature and discharge rate, there will be huge changes to discharging time and capacity fade. From Tables 1–3, Figure 3a–e were constructed. From the figures, the influence of temperature and discharge rate on capacity can be clearly seen. The CE represents the battery performance, while the remaining capacity represents the life of the battery. These two are very important to battery safety. Hence, battery characteristics must be obtained from changes of the CE and remaining capacity under different temperatures, discharge rates, and numbers of cycles to establish a comprehensive battery model to ensure that the battery can operate safely.

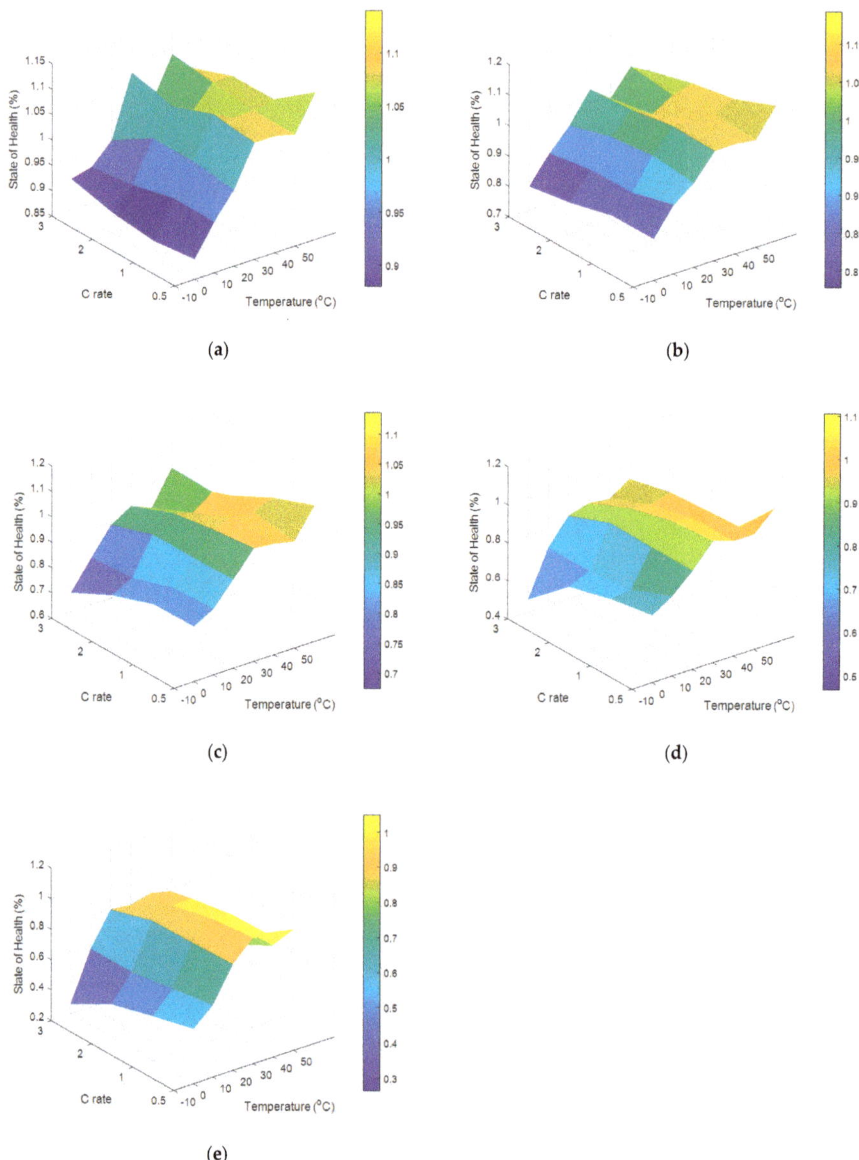

Figure 3. SoH_N at different discharging C rates and temperatures during 400 cycles: (**a**) 1 cycle, (**b**) 100 cycles, (**c**) 200 cycles, (**d**) 300 cycles, and (**e**) 400 cycles.

3. Comprehensive Model Development

3.1. Thevenin Equivalent Circuit Model

The equivalent circuit model is commonly used in predicting battery function and providing estimation to the battery management system [53,54].Figure 4a shows the battery's open-circuit voltage (OCV)-SoC characteristics. Figure 4b shows the battery's voltage-current characteristics. The battery models in Figure 4a,b are quite accurate, since they can clearly describe the battery's nonlinear dynamic behavior, which can be used as a good solution for the SoC estimation of energy storage devices.

The equivalent circuit model can be represented by the following state equations [36,37,53,54], where V_{ps} and V_{pf} are the state variables, I_b is the input, and V_b the output:

$$\dot{V}_{pf} = \frac{1}{R_1 C_1} V_{pf} - \frac{1}{C_1} I_b;$$
$$\dot{V}_{ps} = \frac{1}{R_2 C_2} V_{ps} - \frac{1}{C_2} I_b;$$
$$V_b = V_{oc} + V_{ps} + V_{pf} + R_d I_b,$$

(4)

where V_b is the voltage at the battery terminals, I_b is the current flowing through the battery, V_{oc} is the open circuit voltage, R_b is the internal resistance, R_1 and C_1 comprise the fast resistor–capacitor (RC) network, R_2 and C_2 comprise the slow RC network, V_{pf} is the voltage across the fast RC network, and V_{ps} is the voltage across the slow RC network.

Overall, the equivalent-circuit model is described by the two RC networks shown in Figure 4b. The slow RC network dominates when the battery is nearly full, in which case the discharging behavior is fairly stable. The fast RC network dominates when the battery is running low—during the unstable discharging behavior.

The proposed model is used to estimate the battery's SoC and the remaining capacity by considering the battery model's parameters, and the available states are V_b, I_b, and temperature. Therefore, the simplifications of the electrochemistry equations make it possible to integrate it using equivalent circuit models.

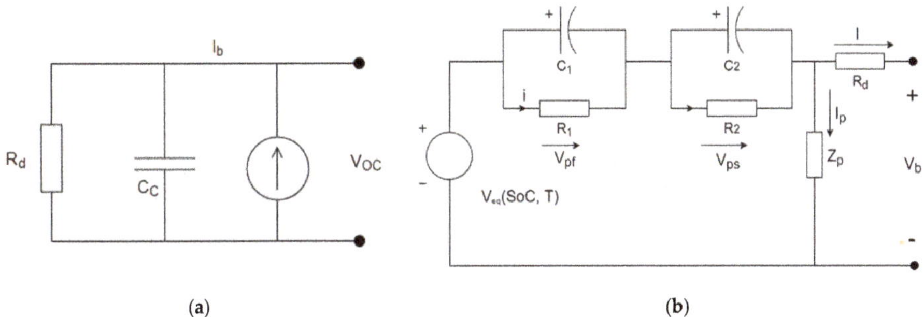

Figure 4. The dual RC equivalent-circuit model for a LiFePO$_4$ battery. (**a**) OCV-SoC characteristics; (**b**) Voltage/current characteristics.

3.2. Butler–Volmer Equation

Capturing of the battery's dynamic responses can be done through the equivalent circuit model descriptions. For the battery's electrochemical reaction phenomena, Butler–Volmer kinetic equation descriptions can be employed, as shown in (5). Charge transfer kinetics of lithium intercalation obeys the Butler–Volmer equation [55–57]. It describes the current from basic electrochemical reactions varying with activation overvoltage η_{act}, such that

$$J = J_0 \cdot \left\{ e^{\left[\frac{\alpha q \eta_{act}}{k_B T} \right]} - e^{\left[\frac{(\alpha - 1) q \eta_{act}}{k_B T} \right]} \right\},$$

(5)

where J is the electrode current density in A/m^2, J_0 is the exchange current density in A/m^2, η_{act} is the surface overpotential in volts, k_B is Boltzmann constant, T is the absolute temperature, $q = 1.602 \times 10^{-19}$(C), α is the anodic charge transfer coefficient, and $(\alpha - 1)$ is the cathodic charge transfer coefficient, where α is assumed to be 0.5 in a battery.

The surface overpotential is as follows: $\eta_{act} = (E - E_{rev})$, where E is the difference of the mean electrostatic potentials of ions and electronsand E_{rev} is the Nernst equilibrium potential, which is strongly related to the local activity of lithium.

Closed circuit voltage (CCV) is defined as $V = V_{ideal} - \eta_{act}$, where V_{ideal} is he standard potential defined by the open circuit voltage plateau ($V_{ideal} = 3.42$ V for Li metal) [5]. The size of η_{act} is dependent on reaction kinetics, meaning J_0 influences η_{act} and it is in turn affected by the reaction temperature. η_{act} represents the voltage loss necessary in overcoming the activation energy barrier of electrochemical reactions. Thus, to obtain a greater current from the battery, a larger voltage loss is inevitable. Changes in electro-chemical reactions can be seen as changes in kinetic parameters, such as α and J_0. If J_0 is low for any specific net current, the kinetics will become sluggish and the activation overvoltage will be even greater. If J_0 is large, a large current can be supplied. In [58], it was stated that since J_0 is difficult to be measured from the outside, Equation (5) can be rewritten as:

$$\frac{1}{J_0 \cdot S} \cdot I(t) = \left[e^{\left[\frac{\alpha q \eta_{act}}{k_B T} \right]} - e^{\left[\frac{(\alpha-1) q \eta_{act}}{k_B T} \right]} \right]; \tag{6}$$

$$I(t) = J_{(t)} \cdot S, \tag{7}$$

where S is the effective area.

In [54], the overvoltage $\eta_{act}(t)$ caused by current variation is represented as:

$$\eta_{act}(t) = \frac{2RT}{F} \ln \left(\frac{1}{2J_0 \cdot S} \cdot I(t) + \sqrt{\left(\frac{1}{2J_0 \cdot S} \cdot I(t) \right)^2 + 1} \right) \tag{8}$$

Although there are still a variety of electrochemical variables that cannot be obtained externally, it is possible to acquire a specific change of potential excited by diverse currents. Thus, it can be seen that J_0 and S are closely related to SoC, reflecting the complexity during the polarization establishment. Increasing the current while fixing other available variables will result in a sharp deviation from the equilibrium state.

3.3. Arrhenius and Peukert Laws

The relationship between the CE and temperature follows the Arrhenius' law [10] in

$$k_a = A \cdot e^{-E_a/RT}; \tag{9}$$

$$Q_{loss} = 1 - \frac{Q}{Q_0}, \tag{10}$$

where k_a is the rate constant, T is the absolute temperature, A is a constant for each chemical reaction known as the pre-exponential factor, E_a is the activation energy for the reaction, R is the universal gas constant, Q_{loss} is the ratio of capacity loss, Q is the actual capacity of the battery, and Q_0 is the battery's initial capacity.

When the battery's discharge current increases, the side effect produced is termed Ohmic heating, meaning part of the energy is wasted as heat. Simultaneously, it also increases Ohmic loss. The rate of ions diffusing and migrating at the poles is lower than the increased discharge current, leading to a lower recovery rate. Finally, it results in a smaller battery capacity. Thus, Peukert's law [23] aims at correcting the relationship between the discharge current and capacity, such that

$$C_t = I^u t, \tag{11}$$

where C_t is the capacity at a one-ampere discharge rate, which must be expressed in ampere hours; I is the actual discharge current in amperes; t is the actual time to discharge the battery; and u is the Peukert constant.

The Peukert constant can be used to evaluate the performance of a battery, like the secondary chemical reactions resulting from battery discharging at a high rate. The materials' structures change as the impedance increases. However, u is still close to 1 [59].

4. Results

4.1. Analysis and Modeling with the Comprehensive Model

From the battery discharging behaviors of Figure 2a–d, fitting curves for the variation of the end of discharging time can be constructed. It is easier to observe the temperature and current dependence of the CE graphically. The curve fitting results are shown in Figure 5, which depicts the difference between measured results and fitted curves. Since the relationship of the CE and temperature in the equivalent circuit model follows the Arrhenius law, while the current obeys Peukert's law, the CE of the equation has the form of:

$$\eta_{c(T)} = a_1 \cdot e^{(b_1 \cdot T)} + a_2 \cdot e^{(b_2 \cdot T)} \tag{12}$$

where T is the temperature, and a_1, b_1, a_2, and b_2 are the fitteing parameters.

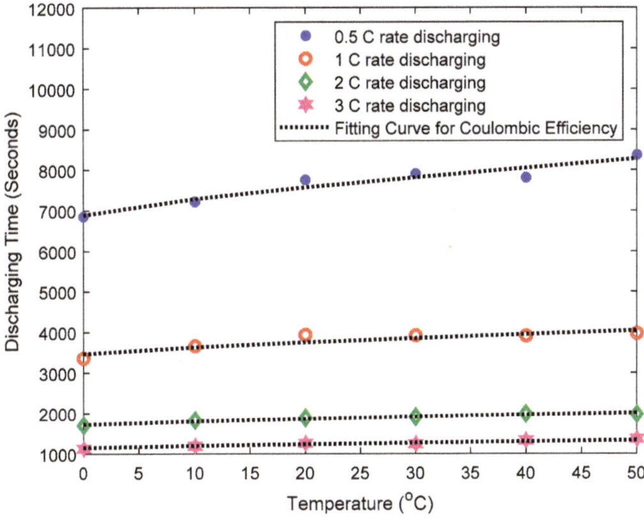

Figure 5. Fitting curves for different discharging temperature and discharging C rates.

The parameters a_1, b_1, a_2, and b_2 of the batteries measured are plotted in Figure 6. It can be clearly seen that the parameters vary with Peukert's law. By observing the trends of the parameters under the same temperature and different discharge currents in Figure 6, the parameters can be fitted to:

$$\begin{aligned}
a_{1(I)} &= p_1 \cdot I^{k_1} \\
b_{1(I)} &= p_2 \cdot I^{k_2} \\
a_{2(I)} &= p_3 \cdot I^{k_3} \\
b_{2(I)} &= p_4 \cdot I^{k_4}
\end{aligned} \tag{13}$$

where I is discharging current; parameters p_1, p_2, p_3, and p_4, k_1, k_2, k_3, and k_4 are fitting parameters. The values are lised in Table 4.

By incorporating all the values of Table 4 for Equation (13) to find the related parameters p_1, p_2, p_3, and p_4, k_1, k_2, k_3, and k_4, we can obtain the predicted values of the parameters a_1, b_1, a_2, and b_2 for Equation (12).

Equation (12) is composed of two exponential terms, which correspond to the pair of RC circuits in Figure 4b, where an exponential term represents the slow RC network; the other exponential term represents the fast RC network.

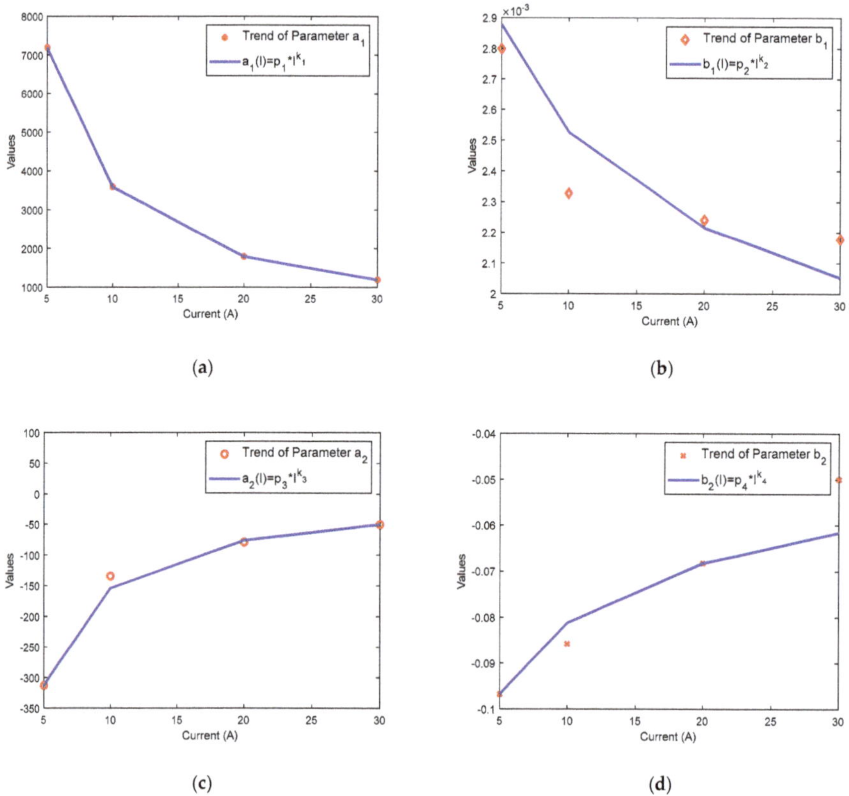

Figure 6. The Peukert law with respect to current can be derived by fitting to obtain the trends of the parameters: (**a**) trend of a_1, (**b**) trend of b_1, (**c**) trend of a_2, and(**d**) trend of b_2.

Comparing the experimental results to Equations (12) and (13), the mean absolute percentage errors (MAPE) defined as

$$MAPE = \frac{\sum_{t=1}^{n} \left| \frac{A_t - F_t}{A_t} \right|}{n} \cdot 100\% \tag{14}$$

are listed in Table 5. In Equation (14), A_t is the measured data of cycle 1 of Tables 1–3, F_t is the fitted data, and n the number of data readings.

Table 4. Parameters are calculated by the columb efficiency (CE) for Equation (12) with Equation (13).

	Parameters			
C Rate	a_1	b_1	a_2	b_2
0.5 C	7.2×10^3	2.799×10^{-3}	-3.130×10^2	-9.669×10^{-2}
1 C	3.6×10^3	2.328×10^{-3}	$-1.3.42 \times 10^2$	-8.570×10^{-2}
2 C	1.8×10^3	2.241×10^{-3}	-7.881×10^1	-6.820×10^{-2}
3 C	1.2×10^3	2.178×10^{-3}	-4.992×10^1	-4.990×10^{-2}

	Parameters							
	p_1	p_2	p_3	p_4	k_1	k_2	k_3	k_4
Value	3.6×10^4	3.9×10^{-3}	-1.628×10^3	-1.45×10^{-1}	-1	-1.887×10^{-1}	-1.025	-2.518×10^{-1}

Table 5. The percentage differences between the experimental and predicted data for all tested batteries (%).

Operating T (°C)		−10	0	10	20	30	40	50
	0.5C	0.3	0.49	1.11	2.21	0.90	3.40	0.60
C rate	1C	2.28	2.59	0.87	4.42	1.15	1.73	2.87
	2C	0.44	1.38	1.52	1.46	0.05	1.31	1.53
	3C	1.33	1.87	0.66	1.69	2.04	2.51	2.75

4.2. Using Artificial Neural Networks to Estimate Remaining Capacity

The CE variation under variable working conditions can be estimated by Equations (12) and (13). When a battery discharges under various environmental conditions during certain cycles, the capacity fades. Without correction, the estimation errors of Equation (1) increase gradually with the operating cycle. Therefore, by introducing the correction of the CE and considering the declination of SoH, the estimation error is effectively reduced.

Considering Equations (2) and (3), $Q_{available}$ represents the available capacity of a battery, which is influenced by the CE. $C_{max,p}$ represents the remaining capacity of a battery, which is not influenced by the CE. In order to obtain the true remaining capacity of a battery, the data of Tables 2–4 were normalized through Equation (15).

$$C_{max,p} = \frac{Q_{available}}{\eta_{c(T)}} \tag{15}$$

The estimation of SoH_N would normally be based on the relationship between charging/discharging cycles and remaining capacity. However, considering actual usage scenarios, such as in electric vehicles and other energy storage devices, the discharge C rate would vary sharply in a short period. Moreover, it is difficult for full charging and discharging to take place. Compound effects under different temperatures and discharge rates on capacity fade are difficult to estimate. Thus, this method cannot be direct applied to scenarios where large variations in environmental parameters exist. Therefore, the charging/discharging cycle numbers of the battery were normalized under different usage environments as discharging times in seconds. Through this method, the accumulated degradation of the batteries under various environments was analyzed. When the battery had undergone various rates of discharge under different temperatures, the remaining capacity was calculated by the ANN.

The learning ability of ANN was realized by means of model training. The common training algorithm is a back-propagation algorithm, which means that the network error will be propagated back if it does not reach an expected value in the model training process; meanwhile, the network weights and biases values are adjusted constantly to obtain the minimum error [60]. This research utilized a multi-layered BPNN to study the model for estimating the remaining capacity ($C_{max,p}$). The network was trained using battery datasets collected under varying temperatures and discharging currents. To build the BPNN model, the complete data was split into training (70%) and testing (30%) data. The detailed operation flow of the BPNN is described below.

The BPNN is one of the most commonly used neural network models. It is a multi-layer feedforward network that can learn and store base on the training of error back propagation. Its rule of learning makes use of gradient descent. Through continual adjustments of back propagation weights and thresholds of the whole network, the network's error square summation (cost) is minimized [60]. Moreover, BPNN can get many input-output models without the knowledge of mathematical equations of the mapping relationship [61].

In experiments, the irreversible capacity fade (C_{fade}) of the battery is influenced by discharging current (i), ambient temperature (T), and cumulative discharging time (cdt). The typical C_{fade} algorithm is as follows:

$$C_{fade} = f(i, T, cdt). \tag{16}$$

Through Equation (16), the topographical structure of BPNN includes: (1) an input layer: input variables, including (T), (i), and (cdt); (2) one or more hidden layers; (3) an output layer. This model is shown graphically in Figure 7.

The parameter ak represents the data input to the model in the Figure 7; tg represents the model output result, which is the remaining capacity of a battery; and w and b are the weight and bias values, respectively. There are three nodes in the input layer. According to Kolmogorov rule [62,63], the number of neurons was set as seven in the hidden layer. The activation function selected for the model construction was log-sigmoid in last two layers [64], written as

$$f(n) = \frac{1}{1 + e^{-n}}. \tag{17}$$

The output value (tg) of each neuron (j) in the hidden layer was calculated by

$$tg_i = f(\sum_{i=1}^{3} ak_i w_{i,j} + b_j),\ i = 1, \ldots, 3,\ j = 1, \ldots, 7, \tag{18}$$

where ak_i represents the input vector, $w_{i,j}$ is the weight value connecting the ith input vector and the jth neuron, b_j represents bias values, and f represents log-sigmoid activation function. The model output tg can be calculated from

$$tg_v = f(\sum_{j=1}^{7} g_j w_{j,v} + b_v),\ v = 1, \tag{19}$$

where v is the number of neurons in the output layer.

In this study, a variant of the gradient descent method called the Levenberg–Marquardt (LM) algorithm was employed to train the constructed model [62,65]. The LM algorithm integrates together, the gradient descent method and the Newton method, which can solve non-linear least squares problems. When it is used for ANN training, the iteration can continue to proceed quickly while ensuring model training speed and accuracy. Finally, the prediction values of remaining capacity were attained finally through the above modeling.

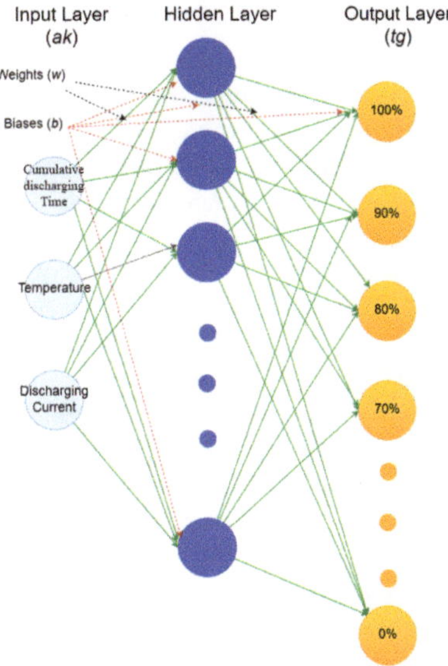

Figure 7. The topographic structure of a back-propagation neural network (BPNN) includes: (1) the input layer: input variables, including (T), (i), and (cdt); (2) one or more hidden layers; (3) an output layer.

4.3. Establishing the Comprehensive Model

This research utilizes the Butler–Volmer equation and the Arrhenius and Peukert laws. In order to express these phenomena in an equivalent circuit model, this research proposed simplification with parameter fitting. In our model, the Butler–Volmer equation describes the voltage and current characteristics of the battery while the temperature and discharging current are modeled by the Arrhenius and Peukert laws. Capacity fade, caused by the compound influence of various operating environments, is calculated by BPNN. Thus, the significance of the equivalent circuit model can be explained as in Figure 8. In the figure, the CE is determined by the two RC networks and the BPNN-modelled capacity fade.

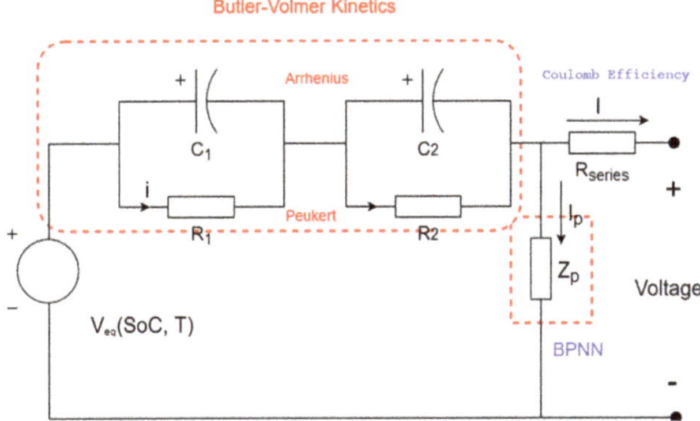

Figure 8. Equivalent-circuit representation for the Butler–Volmer equation, Arrhenius and Peukert laws, and BPNN of a Li-ion battery.

5. Discussions and Verification

Temperature and discharging current are critical parameters in modelling battery dynamics and have been proven to have a significant influence on SoC and remaining capacity estimations. The battery may operate under varying temperatures, including seasonal changes and day/night cycles. The heat generated during the discharge processes will also influence the battery's surface temperatures. In this section, we describe the comprehensive model proposed being tested by the data collected, under varying conditions.

The comprehensive model simultaneously considers the influence of the CE and remaining capacity. In order to verify the correctness of the model, the mean absolute percentage errors (MAPEs) of all tested cases are shown in Figure 9. Overall, the proposed model provides an acceptable estimation result with an MAPE of 1.55% for all the cases tested. As in Figure 9, under room and high temperatures, the proposed model gives a quite satisfying SoC estimation. The estimated SoC are close to the true value, with almost all of the estimation errors being less than 2%. For the SoC estimation under low temperatures, MAPEs vary from 1.06% to 3.71%. It is easily seen from Figure 9d that the estimated SoCs slightly deviate from the true values. As the battery dynamics at low temperature and large C rate are more complicated, SoC estimation is much more difficult. To improve the estimation performance at low temperatures, one practical solution is to increase the amount of training data, particularly for batteries working at low temperatures.

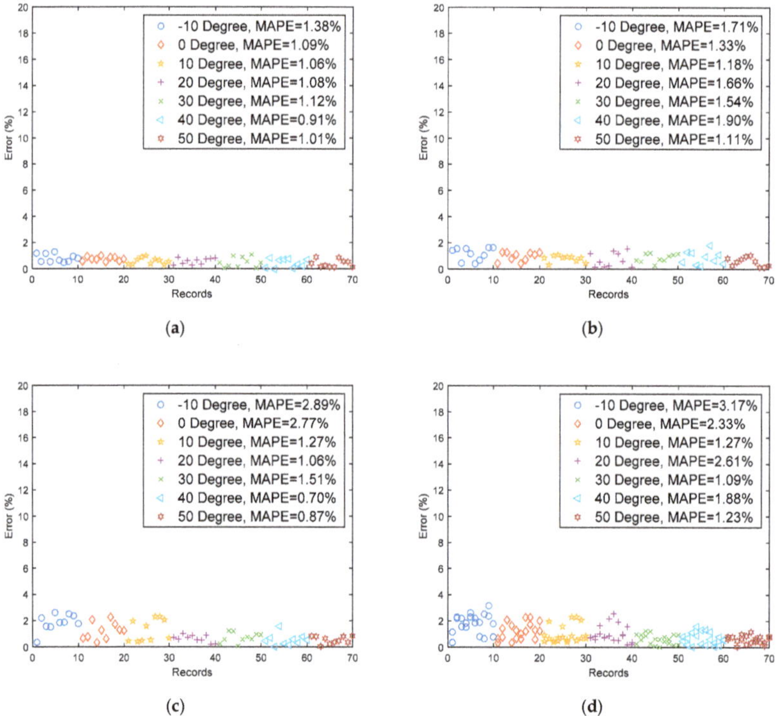

Figure 9. The comparison between the experimental data and simulation results. The experimental data are random samplings in collected records; the simulation results were estimated by the proposed model. (**a**) 0.5 C rate discharge under discharge different temperatures; (**b**) 1 C rate discharge under different temperatures; (**c**) 2 C rate discharge under different temperatures; (**d**) 3 C rate discharge under different temperatures.

6. Conclusions

Coulombic efficiency and remaining capacity are the indicators for estimating battery performance and available capacity, respectively. To ensure safe operation and prevent excessive discharge, accurate SoC and SoH estimations are very important for lithium ion batteries. Thus, much research has been conducted to resolve these issues and increase battery performance. As battery manufacturing technology continues to progress, lithium ion batteries are used more and more. When the battery is discharging at a high rate under a low temperature, the heat produced during discharge will offset some influences from the environmental temperature. As a result, after discharging for a certain length of time, the amount of power delivered will drop while the voltage will conversely rise. In that case, SoC estimates voltage of battery with large errors. As the environmental temperature rises, the battery performance is visibly initially better than that under room temperature or low temperature. However, the battery's cycle life would be lower than that under room temperature. Therefore, under the processes of cyclic charging and discharging, different currents and temperatures produce significant changes in battery capacity. These factors must be considered for the implementation of the battery management system (BMS).

This research used $LiFePO_4$ batteries with 10 Ah capacity and 3.2 V rated voltage for testing under various temperatures and discharge rates. Under testing conditions of 400 cycles, the discharge voltages and currents of the batteries during the discharge process were recorded. The batteries were analyzed with different degrees of aging and their corresponding characteristics were gathered. The comprehensive model was able to follow the temperature and current changes to precisely estimate the SoC of the $LiFePO_4$ battery. The results obtained indicate that the average error of the comprehensive model under various current loads and temperatures to be only 1.55%.

The $LiFePO_4$ batteries are manufactured in the same factory and should have a high degree of consistency. Therefore, when the error of one battery is higher than the rest, it can be considered a signal that the battery capacity is fading faster than before, indicating that the battery should be replaced immediately. Due to the model accuracy and stability shown during this paper, the model can be considered a useful tool for the control and performance analysis of a management system which includes a Li-ion battery.

Funding: This research received no external funding.

Conflicts of Interest: The author declares no conflict of interest.

References

1. Kurzweil, P.; Brandt, K. Secondary batteries–lithium rechargeable systems| Overview. In *Encyclopedia of Electrochemical Power Sources*; Newnes: Oxford, UK, 2009; pp. 1–26.
2. Garche, J.; Dyer, C.K.; Moseley, P.T.; Ogumi, Z.; Rand, D.A.; Scrosati, B. *Encyclopedia of Electrochemical Power Sources*; Newnes: Oxford, UK, 2013.
3. Ali, M.U.; Zafar, A.; Nengroo, S.H.; Hussain, S.; Alvi, M.J.; Kim, H.J.J.E. Towards a Smarter Battery Management System for Electric Vehicle Applications: A Critical Review of Lithium-Ion Battery State of Charge Estimation. *Energies* **2019**, *12*, 446. [CrossRef]
4. Miao, Y.; Hynan, P.; von Jouanne, A.; Yokochi, A.J.E. Current Li-Ion Battery Technologies in Electric Vehicles and Opportunities for Advancements. *Energies* **2019**, *12*, 1074. [CrossRef]
5. Bai, P.; Cogswell, D.A.; Bazant, M.Z. Suppression of phase separation in LiFePO(4) nanoparticles during battery discharge. *Nano Lett.* **2011**, *11*, 4890–4896. [CrossRef] [PubMed]
6. Chan, H.L. A new battery model for use with battery energy storage systems and electric vehicles power systems. In Proceedings of the 2000 IEEE Power Engineering Society Winter Meeting. Conference Proceedings (Cat. No.00CH37077), Singapore, 23–27 January 2000; Volume 1, pp. 470–475. [CrossRef]
7. Chen, M.; Rincon-Mora, G.A. Accurate Electrical Battery Model Capable of Predicting Runtime and I–V Performance. *IEEE Trans. Energy Convers.* **2006**, *21*, 504–511. [CrossRef]

8. Dargaville, S.; Farrell, T.W. Predicting Active Material Utilization in LiFePO4 Electrodes Using a Multiscale Mathematical Model. *J. Electrochem. Soc.* **2010**, *157*, A830–A840. [CrossRef]

9. Kuo, T.J.; Lee, K.Y.; Huang, C.K.; Chen, J.H.; Chiu, W.L.; Huang, C.F.; Wu, S.D. State of charge modeling of lithium-ion batteries using dual exponential functions. *J. Power Sources* **2016**, *315*, 331–338. [CrossRef]

10. Sun, S.; Guan, T.; Shen, B.; Leng, K.; Gao, Y.; Cheng, X.; Yin, G. Changes of Degradation Mechanisms of LiFePO 4 /Graphite Batteries Cycled at Different Ambient Temperatures. *Electrochim. Acta* **2017**, *237*, 248–258. [CrossRef]

11. Wang, J.; Yang, J.; Tang, Y.; Li, R.; Liang, G.; Sham, T.K.; Sun, X. Surface aging at olivine LiFePO4: A direct visual observation of iron dissolution and the protection role of nano-carbon coating. *J. Mater. Chem. A* **2013**, *1*, 1579–1586. [CrossRef]

12. Chang, W.; Kim, S.J.; Park, I.T.; Cho, B.W.; Chung, K.Y.; Shin, H.C. Low temperature performance of LiFePO4 cathode material for Li-ion batteries. *J. Alloys Compd.* **2013**, *563*, 249–253. [CrossRef]

13. Zhang, S.S.; Xu, K.; Jow, T.R. A new approach toward improved low temperature performance of Li-ion battery. *Electrochem. Commun.* **2002**, *4*, 928–932. [CrossRef]

14. Abraham, D.P.; Heaton, J.R.; Kang, S.H.; Dees, D.W.; Jansen, A.N. Investigating the Low-Temperature Impedance Increase of Lithium-Ion Cells. *J. Electrochem. Soc.* **2008**, *155*, A41. [CrossRef]

15. Maheshwari, A.; Heck, M.; Santarelli, M. Cycle aging studies of lithium nickel manganese cobalt oxide-based batteries using electrochemical impedance spectroscopy. *Electrochim. Acta* **2018**, *273*, 335–348. [CrossRef]

16. Anseán, D.; Dubarry, M.; Devie, A.; Liaw, B.Y.; García, V.M.; Viera, J.C.; González, M. Operando lithium plating quantification and early detection of a commercial LiFePO4 cell cycled under dynamic driving schedule. *J. Power Sources* **2017**, *356*, 36–46. [CrossRef]

17. Waldmann, T.; Wilka, M.; Kasper, M.; Fleischhammer, M.; Wohlfahrt-Mehrens, M. Temperature dependent ageing mechanisms in Lithium-ion batteries—A Post-Mortem study. *J. Power Sources* **2014**, *262*, 129–135. [CrossRef]

18. Agubra, V.; Fergus, J.J.M. Lithium ion battery anode aging mechanisms. *Materials* **2013**, *6*, 1310–1325. [CrossRef]

19. Hendricks, C.; Williard, N.; Mathew, S.; Pecht, M. A failure modes, mechanisms, and effects analysis (FMMEA) of lithium-ion batteries. *J. Power Sources* **2015**, *297*, 113–120. [CrossRef]

20. Song, W.; Chen, M.; Bai, F.; Lin, S.; Chen, Y.; Feng, Z. Non-uniform effect on the thermal/aging performance of Lithium-ion pouch battery. *Appl. Therm. Eng.* **2018**, *128*, 1165–1174. [CrossRef]

21. Yao, X.Y.; Pecht, M.G. Tab Design and Failures in Cylindrical Li-ion Batteries. *IEEE Access* **2019**, *7*, 24082–24095. [CrossRef]

22. Ko, S.T.; Lee, J.; Ahn, J.H.; Lee, B.K.J.E. Innovative Modeling Approach for Li-Ion Battery Packs Considering Intrinsic Cell Unbalances and Packaging Elements. *Energies* **2019**, *12*, 356. [CrossRef]

23. Li, J.; Zhang, J.; Zhang, X.; Yang, C.; Xu, N.; Xia, B. Study of the storage performance of a Li-ion cell at elevated temperature. *Electrochim. Acta* **2010**, *55*, 927–934. [CrossRef]

24. Guan, T.; Sun, S.; Gao, Y.; Du, C.; Zuo, P.; Cui, Y.; Zhang, L.; Yin, G. The effect of elevated temperature on the accelerated aging of LiCoO 2 /mesocarbon microbeads batteries. *Appl. Energy* **2016**, *177*, 1–10. [CrossRef]

25. Kassem, M.; Delacourt, C. Postmortem analysis of calendar-aged graphite/LiFePO4 cells. *J. Power Sources* **2013**, *235*, 159–171. [CrossRef]

26. Smith, A.J.; Burns, J.C.; Trussler, S.; Dahn, J.R. Precision Measurements of the Coulombic Efficiency of Lithium-Ion Batteries and of Electrode Materials for Lithium-Ion Batteries. *J. Electrochem. Soc.* **2010**, *157*, A196–A202. [CrossRef]

27. Bond, T.M.; Burns, J.C.; Stevens, D.A.; Dahn, H.M.; Dahn, J.R. Improving Precision and Accuracy in Coulombic Efficiency Measurements of Li-Ion Batteries. *J. Electrochem. Soc.* **2013**, *160*, A521–A527. [CrossRef]

28. Santhanagopalan, S.; White, R.E. Online estimation of the state of charge of a lithium ion cell. *J. Power Sources* **2006**, *161*, 1346–1355. [CrossRef]

29. Huang, S.C.; Tseng, K.H.; Liang, J.W.; Chang, C.L.; Pecht, M.G. An Online SOC and SOH Estimation Model for Lithium-Ion Batteries. *Energies* **2017**, *10*, 512. [CrossRef]

30. Berrueta, A.; Urtasun, A.; Ursúa, A.; Sanchis, P. A comprehensive model for lithium-ion batteries: From the physical principles to an electrical model. *Energy* **2018**, *144*, 286–300. [CrossRef]

31. Tippmann, S.; Walper, D.; Balboa, L.; Spier, B.; Bessler, W.G. Low-temperature charging of lithium-ion cells part I: Electrochemical modeling and experimental investigation of degradation behavior. *J. Power Sources* **2014**, *252*, 305–316. [CrossRef]

32. Salvadori, A.; Grazioli, D.; Geers, M.G.D. Governing equations for a two-scale analysis of Li-ion battery cells. *Int. J. Solids Struct* **2015**, *59*, 90–109. [CrossRef]

33. Gu, L.; Xiao, D.; Hu, Y.S.; Li, H.; Ikuhara, Y. Atomic-scale structure evolution in a quasi-equilibrated electrochemical process of electrode materials for rechargeable batteries. *Adv. Mater.* **2015**, *27*, 2134–2149. [CrossRef]

34. Fuller, T.F.; Doyle, M.; Newman, J. Relaxation Phenomena in Lithium-Ion-Insertion Cells. *J. Electrochem. Soc.* **1994**, *141*, 982–990. [CrossRef]

35. Singh, M.; Kaiser, J.; Hahn, H. Effect of Porosity on the Thick Electrodes for High Energy Density Lithium Ion Batteries for Stationary Applications. *Batteries* **2016**, *2*, 35. [CrossRef]

36. Petzl, M.; Danzer, M.A. Advancements in OCV Measurement and Analysis for Lithium-Ion Batteries. *IEEE Trans. Energy Convers.* **2013**, *28*, 675–681. [CrossRef]

37. Chaoui, H.; Ibe-Ekeocha, C.C.; Gualous, H. Aging prediction and state of charge estimation of a LiFePO 4 battery using input time-delayed neural networks. *Electr. Power Syst. Res.* **2017**, *146*, 189–197. [CrossRef]

38. Khumprom, P.; Yodo, N.J.E. A Data-Driven Predictive Prognostic Model for Lithium-Ion Batteries based on a Deep Learning Algorithm. *Energies* **2019**, *12*, 660. [CrossRef]

39. Wang, Y.J.; Yang, D.; Zhang, X.; Chen, Z.H. Probability based remaining capacity estimation using data-driven and neural network model. *J. Power Sources* **2016**, *315*, 199–208. [CrossRef]

40. Dong, G.Z.; Zhang, X.; Zhang, C.B.; Chen, Z.H. A method for state of energy estimation of lithium-ion batteries based on neural network model. *Energy* **2015**, *90*, 879–888. [CrossRef]

41. Hu, X.S.; Li, S.B.; Peng, H. A comparative study of equivalent circuit models for Li-ion batteries. *J. Power Sources* **2012**, *198*, 359–367. [CrossRef]

42. Wang, Q.K.; He, Y.J.; Shen, J.N.; Ma, Z.F.; Zhong, G.B. A unified modeling framework for lithium-ion batteries: An artificial neural network based thermal coupled equivalent circuit model approach. *Energy* **2017**, *138*, 118–132. [CrossRef]

43. Jokar, A.; Désilets, M.; Lacroix, M.; Zaghib, K. Mesoscopic modeling and parameter estimation of a lithium-ion battery based on LiFePO4/graphite. *J. Power Sources* **2018**, *379*, 84–90. [CrossRef]

44. Saxena, S.; Kang, M.; Xing, Y.; Pecht, M. Anomaly Detection During Lithium-ion Battery Qualification Testing. In Proceedings of the 2018 IEEE International Conference on Prognostics and Health Management (ICPHM), Seattle, WA, USA, 11–13 June 2018; pp. 1–6.

45. Diao, W.; Saxena, S.; Han, B.; Pecht, M. Algorithm to Determine the Knee Point on Capacity Fade Curves of Lithium-Ion Cells. *Energies* **2019**, *12*. [CrossRef]

46. Razavi-Far, R.; Farajzadeh-Zanjani, M.; Chakrabarti, S.; Saif, M. Data-driven prognostic techniques for estimation of the remaining useful life of lithium-ion batteries. In Proceedings of the 2016 IEEE International Conference on Prognostics and Health Management (ICPHM), Ottawa, ON, Canada, 20–22 June 2016; pp. 1–8.

47. Jiang, J.; Gao, Y.; Zhang, C.; Zhang, W.; Jiang, Y. Lifetime Rapid Evaluation Method for Lithium-Ion Battery with Li(NiMnCo)O2 Cathode. *J. Electrochem. Soc.* **2019**, *166*, A1070–A1081. [CrossRef]

48. Kennedy, B.; Patterson, D.; Camilleri, S. Use of lithium-ion batteries in electric vehicles. *J. Power Sources* **2000**, *90*, 156–162. [CrossRef]

49. Zhang, T.; Fuchs, B.; Secchiaroli, M.; Wohlfahrt-Mehrens, M.; Dsoke, S. Electrochemical behavior and stability of a commercial activated carbon in various organic electrolyte combinations containing Li-salts. *Electrochim. Acta* **2016**, *218*, 163–173. [CrossRef]

50. Zhang, Y.C.; Wang, C.Y.; Tang, X.D. Cycling degradation of an automotive LiFePO4 lithium-ion battery. *J. Power Sources* **2011**, *196*, 1513–1520. [CrossRef]

51. Wang, J.; Liu, P.; Hicks-Garner, J.; Sherman, E.; Soukiazian, S.; Verbrugge, M.; Tataria, H.; Musser, J.; Finamore, P. Cycle-life model for graphite-LiFePO4 cells. *J. Power Sources* **2011**, *196*, 3942–3948. [CrossRef]

52. Shim, J.; Striebel, K.A. Cycling performance of low-cost lithium ion batteries with natural graphite and LiFePO4. *J. Power Sources* **2003**, *119*, 955–958. [CrossRef]

53. Tran, N.T.; Khan, A.; Nguyen, T.T.; Kim, D.W.; Choi, W.J.E. SOC Estimation of Multiple Lithium-Ion Battery Cells in a Module Using a Nonlinear State Observer and Online Parameter Estimation. *Energies* **2018**, *11*, 1620. [CrossRef]

54. Huangfu, Y.; Xu, J.; Zhao, D.; Liu, Y.; Gao, F.J.E. A novel battery state of charge estimation method based on a super-twisting sliding mode observer. *Energies* **2018**, *11*, 1211. [CrossRef]

55. Heubner, C.; Schneider, M.; Michaelis, A. Investigation of charge transfer kinetics of Li-Intercalation in LiFePO$_4$. *J. Power Sources* **2015**, *288*, 115–120. [CrossRef]

56. Bai, P.; Bazant, M.Z. Charge transfer kinetics at the solid-solid interface in porous electrodes. *Nat. Commun.* **2014**, *5*. [CrossRef] [PubMed]

57. Hess, A.; Roode-Gutzmer, Q.; Heubner, C.; Schneider, M.; Michaelis, A.; Bobeth, M.; Cuniberti, G. Determination of state of charge-dependent asymmetric Butler–Volmer kinetics for LixCoO2 electrode using GITT measurements. *J. Power Sources* **2015**, *299*, 156–161. [CrossRef]

58. Liu, S.; Jiang, J.; Shi, W.; Ma, Z.; Wang, L.Y.; Guo, H. Butler–Volmer-Equation-Based Electrical Model for High-Power Lithium Titanate Batteries Used in Electric Vehicles. *IEEE Trans. Ind. Electron.* **2015**, *62*, 7557–7568. [CrossRef]

59. O'Malley, R.; Liu, L.; Depcik, C. Comparative study of various cathodes for lithium ion batteries using an enhanced Peukert capacity model. *J. Power Sources* **2018**, *396*, 621–631. [CrossRef]

60. Kalogirou, S. Artificial neural networks for the prediction of the energy consumption of a passive solar building. *Energy* **2000**, *25*, 479–491. [CrossRef]

61. Golden, R.M.; Golden, R. *Mathematical Methods for Neural Network Analysis and Design*; MIT Press: Cambridge, MA, USA, 1996.

62. Fang, K.; Mu, D.; Chen, S.; Wu, B.; Wu, F. A prediction model based on artificial neural network for surface temperature simulation of nickel–metal hydride battery during charging. *J. Power Sources* **2012**, *208*, 378–382. [CrossRef]

63. Miller, W.T.; Werbos, P.J.; Sutton, R.S. *Neural Networks for Control*; MIT press: Cambridge, MA, USA, 1995.

64. Hannan, M.A.; Lipu, M.S.H.; Hussain, A.; Saad, M.H.; Ayob, A. Neural Network Approach for Estimating State of Charge of Lithium-Ion Battery Using Backtracking Search Algorithm. *IEEE Access* **2018**, *6*, 10069–10079. [CrossRef]

65. Haykin, S. *Neural Networks: A Comprehensive Foundation*; Prentice Hall PTR: Upper sardeliver, NJ, USA, 1994.

Article

Small-Signal Modeling and Analysis for a Wirelessly Distributed and Enabled Battery Energy Storage System of Electric Vehicles

Yuan Cao

Department of Electrical and Computer Engineering, College of Engineering, The University of Alabama, Tuscaloosa, AL 35487, USA; ycao19@crimson.ua.edu

Received: 31 August 2019; Accepted: 7 October 2019; Published: 11 October 2019

Featured Application: The major object of this research is to provide a small-signal modeling method and controller design guidelines in wireless distributed and enabled battery energy storage system (WEDES) battery system for electric vehicles applications.

Abstract: This paper presents small-signal modeling, analysis, and control design for wireless distributed and enabled battery energy storage system (WEDES) for electric vehicles (EVs), which can realize the active state-of-charge (SOC) balancing between each WEDES battery module and maintain operation with a regulated bus voltage. The derived small-signal models of the WEDES system consist of several sub-models, such as the DC-DC boost converter model, wireless power transfer model, and the models of control compensators. The small-signal models are able to provide deep insight analysis of the steady-state and dynamics of the WEDES battery system and provide design guidelines or criteria of the WEDES controller. The derived small-signal models and controller design are evaluated and validated by both MATLAB®/SIMULINK simulation and hardware experimental prototype.

Keywords: small-signal modeling; battery energy storage system; battery management system; control; stability; dynamic response; wireless power; state-of-charge; electric vehicle

---

1. Introduction

Battery energy storage systems (BESS) have been widely used in various applications, such as electric vehicles (EVs), consumer electronics, medical devices, smart grid, energy backup in data centers, and among others [1–10]. For EV applications, what is referred to be as "range anxiety" is one of the major reasons that prohibit or slows down the adoption of EVs [9–14].

To eliminate range anxiety in and extend the driving range of EVs, different methods have been discussed in the literature [9–17], such as increasing battery pack capacity, utilizing a faster charging method, utilizing a battery pack swapping method, achieving dynamic wireless charging, etc. While these methods can be effective to extend the driving range of EVs, some design challenges or drawbacks cannot be ignored.

When increasing capacity, the weight, size, and cost of the battery pack inevitably increase with the increase in battery capacity [10]. Further, the needed recharge time is also increased. For faster charging, the battery state-of-health (SOH) degrades at a higher rate if faster charging is applied [11,12]. In addition, the fast charger requires a high-power infrastructure that increases the cost of the overall system. For conventional battery swapping, specialized equipment, as well as the experienced personnel, are required to realize battery swapping [13]. For dynamic wireless charging, a large number of transmitter (Tx) coils are required with corresponding power supply units, which increases the infrastructure cost. In addition, this method might not be practical in all locations [14].

Among these methods, the battery swapping concept is a good candidate to reduce recharging time and extend the driving range with low infrastructure cost. To deal with the challenges in the conventional battery swapping concept, a new distributed and enabled battery energy storage (WEDES) system and WEDES controller for EVs are presented in [1], which allows for fast and safe exchange/swapping of smaller and lighter battery modules with wireless power transfer (WPT) technology [18,19].

An illustration of the WEDES system for EVs is shown in Figure 1b, and its example circuit diagram is shown in Figure 1c. Each of the battery modules consists of multiple battery cells, a dedicated electronics circuit, wireless power transmitter coil (Tx coil), wireless communication circuit, and client controller. While the on-board-unit (OBU) consists of a wireless power receiver coil (Rx coil), wireless communication circuit, and host controller. Different from the conventional battery swapping concept where the battery pack as a whole is exchanged at one time, in the WEDES battery system, the conventional single battery pack is divided into multiple small battery modules, which can deliver power through wireless power transfer (WPT) technology to the OBU. The distributed nature of the WEDES system combined with wireless power transfer (no physical connection between battery modules and OBU) makes the battery exchange/swapping easier, safer, and faster.

The distributed WEDES battery system with the WEDES controller addresses state-of-charge (SOC) balancing, bus voltage regulation, and battery module current/voltage regulation at the same time inside the system. Therefore, an SOC balancing control loop, a bus voltage regulation control loop, and a battery module current/voltage control loop are coupled with each other within one battery module as well as between multiple battery modules. These couplings make the analysis and design of the WEDES controller complex and critical. While the initial concept of the distributed WEDES battery system is discussed in [1], the small-signal modeling and controller design analysis are not focused on.

The main contributions of this paper can be summarized as follows:

(1). The overall review of the wireless distributed WEDES system, which allows for fast and safe exchange/swapping of battery modules when utilized in electric vehicles (EVs) applications to deal with the range anxiety issue.
(2). The derivation of small-signal modeling of the WEDES battery system to comprehensively analyze the steady-state stability and dynamic response of the WEDES battery system.
(3). The discussion of the guidelines for the controller design of multiple interacted control loops.
(4). The discussion of the simulation results and hardware experimental results to evaluate and validate the accuracy and effectiveness of the derived small-signal model and designed compensators.

The next Section discusses the detailed small-signal derivation of the WEDES system. Section 3 presents the design of compensators for each control loop. Simulation models and experimental results are presented and discussed in Section 4 to validate the derived small-signal models. Section 5 is the additional comments, and Section 6 concludes the paper.

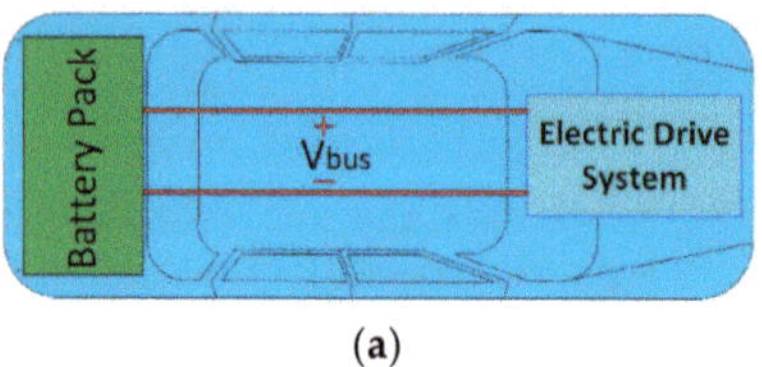

(a)

Figure 1. *Cont.*

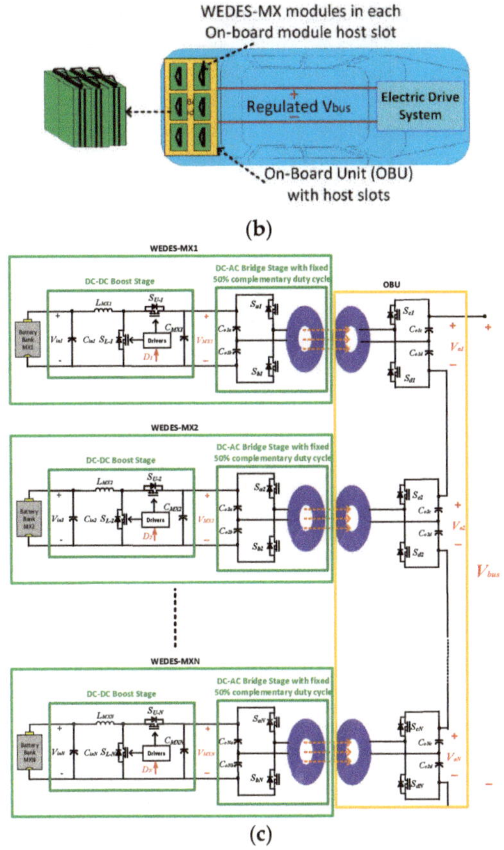

Figure 1. Illustration diagrams of battery system for electric vehicle (EV) application. (**a**) The conventional battery pack and electrics drive system in EVs, (**b**) the wireless distributed and enabled battery energy storage (WEDES) battery system in EVs, and (**c**) example circuit diagram of the WEDES system [1].

2. Small-Signal Modeling of the Distributed WEDES System

2.1. Overview of the WEDES System and Controller Operation Principle

Figure 1c shows the illustration of an example circuit diagram of the WEDES system, which consists of two major parts: battery modules and on-board-unit (OBU).

Inside each battery module, multiple battery cells are connected in series and/or in parallel to form a battery bank, which can provide voltage/current/power to the rest of the system. The output of the battery bank is then connected to the input of a DC-DC power converter, which is used to achieve bus voltage regulation, battery module current/voltage regulation as well as SOC balancing at the same time as described later in this section. The output of the power converter is connected to an inverter stage for DC-AC power conversion. At the end of the battery module, the AC power from the inverter is applied to the transmitter (Tx) for inductive wireless power transfer (I-WPT) to the OBU.

The OBU mainly consists of receiver coils (Rx) followed by an AC-DC power conversion/rectification stage (rectifier). The outputs of each rectifier (V_{o1} through V_{oN}) are connected in series to the bus/output ($V_{bus} = V_{o1} + V_{o2} + \ldots + V_{oN}$).

To realize the functionalities of SOC balancing, bus voltage regulation, and battery module current/voltage regulation, the WEDES controller consists of three different control loops: the SOC balancing control loop (referred to be by the SOC balancing loop), battery module voltage control loop (referred to be by the module voltage loop) and bus voltage control loop (referred to be by the bus voltage loop). Figure 2 shows the diagram of the wirelessly distributed WEDES controller, where V_{bus_ref} is the desired value of bus voltage, $V_{MN\text{-}total}$ is an intermediate value for voltage regulation, $C_{\_re\_MX1}$ though $C_{\_re\_MXN}$ are the remaining capacities of battery modules for SOC calculation, SOC_{MX1} though SOC_{MXN} are the SOC values of battery modules, λ_{DC1} through λ_{DCN} are weighting factors to generate the reference values of $V_{MX1\text{-}DC\text{-}ref}$ through $V_{MXN\text{-}DC\text{-}ref}$ for each battery module, α_{MX1} through α_{MXN} are the SOC multipliers and δ_{MX1} through δ_{MXN} are enable/disable values.

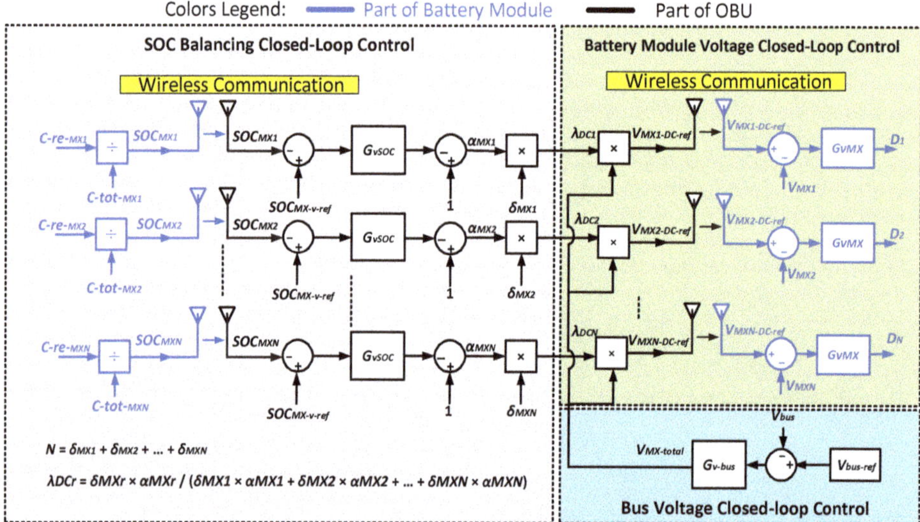

Figure 2. Illustration of the WEDES controller presented in [1].

In the WEDES controller, the SOC balancing loop is used to generate multipliers α_{MX1} through α_{MXN} to realize SOC balancing between multiple battery modules, as given by Equation (1).

$$\begin{cases} \alpha_{MX1} = (SOC_{MX-ref} - SOC_{MX1}) \times G_{vSOC} + 1 \\ \alpha_{MX2} = (SOC_{MX-ref} - SOC_{MX2}) \times G_{vSOC} + 1 \\ \qquad \cdots\cdots \\ \alpha_{MXN} = (SOC_{MX-ref} - SOC_{MXN}) \times G_{vSOC} + 1 \end{cases}, \tag{1}$$

where $SOC_{MX\text{-}ref}$ is the average SOC value of all battery modules, as given by Equation (2). When all battery modules are inserted and active, the sum of all δ_{MX1} through δ_{MXN} equals to N ($\delta_{MX1} + \delta_{MX2} + \ldots + \delta_{MXN} = N$).

$$SOC_{MX-ref} = \frac{(\delta_{MX1} \times SOC_{MX1}) + (\delta_{MX2} \times SOC_{MX2}) + \ldots + (\delta_{MXN} \times SOC_{MXN})}{\delta_{MX1} + \delta_{MX2} + \ldots + \delta_{MXN}}, \tag{2}$$

$$\lambda_{DCr} = \frac{\delta_{MXr} \times \alpha_{MXr}}{(\delta_{MX1} \times \alpha_{MX1}) + (\delta_{MX2} \times \alpha_{MX2}) + \ldots + (\delta_{MXN} \times \alpha_{MXN})} \tag{3}$$

If the SOC value of rth battery module is larger than others, the corresponding multiplier α_{MXr} will be set larger, and vice versa. These multipliers α_{MX1} through α_{MXN} are then multiplied by enabled/disable values δ_{MX1} through δ_{MXN} to further generate the weighting factors λ_{DC1} through λ_{DCN}, as given by Equation (3). The sum of weighting factors λ_{DC1} through λ_{DCN} always equals to one. These weighting factors are then used in the battery module voltage loop to regulate the output voltage of the battery modules V_{MX1} through V_{MXN} at the primary side (Tx side), and as a result, achieve bus voltage regulation at the second side (Rx/OBU side).

It should be emphasized that due to the inevitable power loss during wireless power transfer (i.e., transmission efficiency is less than 100%), the bus voltage control loop is important to adaptively adjust $V_{MN\text{-total}}$ value to compensate the conversion ratios and losses of multiple power conversion stages (DC-AC-AC-DC) and realize bus voltage regulation. The relationship between different voltages can be calculated as given by Equation (4).

$$
\begin{cases}
V_{MX-total} = \left(V_{bus-ref} - V_{bus}\right) \times G_{v-bus} \\
V_{MX-total} = V_{MX1} + V_{MX2} + \ldots + V_{MXN} \\
V_{MXr-DC-ref} = V_{MX-total} \times \lambda_{DCr} \\
V_{bus} = V_{o1} + V_{o2} + \ldots + V_{oN}
\end{cases}
\tag{4}
$$

To summarize, the presented WEDES controller can dynamically control SOC multipliers α_{MX1} through α_{MXN} to adjust the discharging rate for each battery module to achieve SOC balancing, while keeping $\lambda_{DC1} + \lambda_{DC2+} \ldots + \lambda_{DCN} = 1$ such that the bus voltage is always regulated as $V_{bus\text{-ref}}$.

2.2. Small-signal Modeling

Based on the block diagram of the WEDES controller shown in Figure 2, the small-signal of the distributed WEDES system with controller is shown in Figure 3. The transfer functions and symbols in Figure 3 are summarized as follows. For simplicity, the rth battery module is used for illustration.

$L_{bus}(s)$: Bus voltage control loop gain;

$L_{MXr}(s)$: Battery module voltage control loop gain;

$L_{SOCr}(s)$: SOC balancing control loop gain;

$G_{vdr}(s)$: Duty cycle to DC-DC converter output voltage V_{MXN} transfer function;

$G_{idN}(s)$: Duty cycle to cell current transfer function;

$G_{socir}(s)$: Cell current to cell SOC transfer function;

$G_{iTR}(s)$: Gain of the input current of the half-bridge inverter to the output current of rectifier;

$G_{vTR}(s)$: Gain of the input voltage of the half-bridge inverter to the output voltage of rectifier;

G_{PWM}: PWM module gain;

$K_{divr}(s)$: Output voltage sensing gain (including voltage divider gain and ADC conversion gain);

$Delay_{wr}(s)$: Delay of wireless communication;

$Delay_{dr}(s)$: Delay of digital computation;

$ZOH_{vr}(s)$: Zero order hold model of voltage sampling;

$ZOH_{ir}(s)$: Zero order hold model of current sampling;

$ZOH_{SOCr}(s)$: Zero order hold model of SOC sampling.

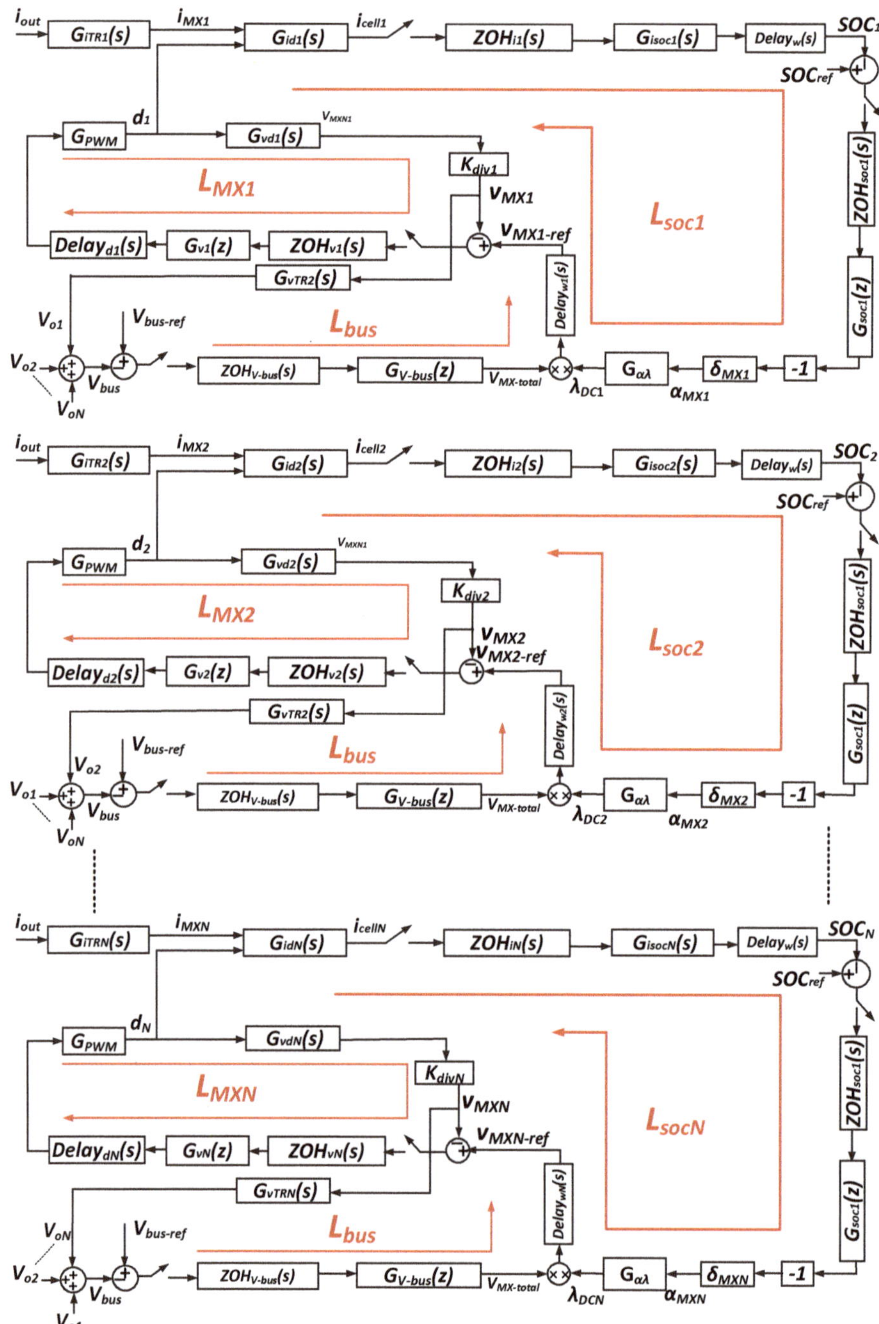

Figure 3. Small-signal of distributed the WEDES battery system with the WEDES controller.

2.3. Derivation of Transfer Functions

Since the design parameters and equilibrium operation point of all WEDES battery modules are the same under steady-state operation (when SOC balancing is achieved), the derivation of transfer functions of all battery modules follows the same procedure. The detailed derivation of rth battery module is discussed as follows:

2.3.1. Transfer Function of DC-DC Boost Stage

The circuit diagram of the DC-DC boost stage is shown in Figure 4. When the lower side switch $S_{L\text{-}r}$ is on and the upper side switch $S_{U\text{-}r}$ is off, the differential equation of the boost converter can be derived as follows:

$$\begin{cases} L_{MXr}\frac{di_{inr}}{dt} = v_{inr} \\ C_{MXr}\frac{dv_{MXr}}{dt} = -i_{MXr} \end{cases},$$

(5)

where i_{inr} and v_{inr} are the input current and input voltage, respectively, and i_{MXr} and v_{MXr} are the output current and output voltage of the boost converter, respectively. L_{MXr} is the inductor value, and C_{MXr} is the output capacitor.

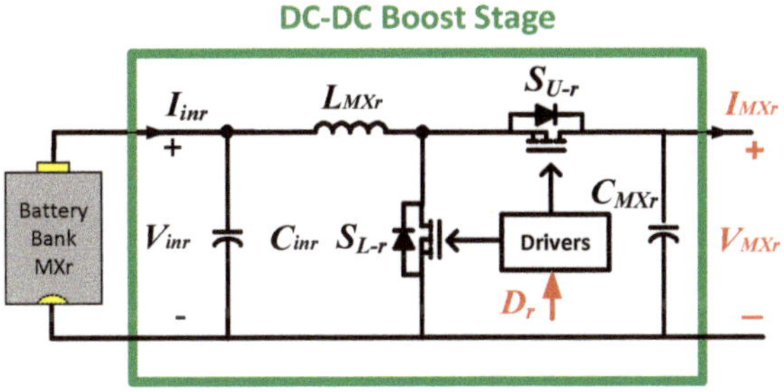

Figure 4. Circuit diagram of DC-DC boost stage in the WEDES system.

The state-space form of Equation (5) can be rewritten as Equation (6).

$$K\begin{bmatrix} \dot{x}_1 \\ \dot{x}_2 \end{bmatrix} = A_1\begin{bmatrix} x_1 \\ x_2 \end{bmatrix} + B_1\begin{bmatrix} u_1 \\ u_2 \end{bmatrix} = \begin{bmatrix} 0 & 0 \\ 0 & 0 \end{bmatrix}\begin{bmatrix} x_1 \\ x_2 \end{bmatrix} + \begin{bmatrix} 1 & 0 \\ 0 & -1 \end{bmatrix}\begin{bmatrix} u_1 \\ u_2 \end{bmatrix},$$

(6)

where the variables are $x_1 = i_{inr}$, $x_2 = v_{MXr}$, $u_1 = v_{inr}$, $u_2 = i_{MXr}$, respectively. $K = \begin{bmatrix} L & 0 \\ 0 & C \end{bmatrix}$.

Similarly, when the lower side switch $S_{L\text{-}r}$ is off and the upper side switch $S_{U\text{-}r}$ is on, the differential equation of the boost converter can be derived as follows:

$$\begin{cases} L_{MXr}\frac{di_{inr}}{dt} = V_{inr} - V_{MXr} \\ C_{MXr}\frac{dV_{MXr}}{dt} = i_{inr} - I_{MXr} \end{cases}.$$

(7)

The state-space form of Equation (7) can be rewritten as Equation (8).

$$K\begin{bmatrix} \dot{x}_1 \\ \dot{x}_2 \end{bmatrix} = A_2\begin{bmatrix} x_1 \\ x_2 \end{bmatrix} + B_2\begin{bmatrix} u_1 \\ u_2 \end{bmatrix} = \begin{bmatrix} 0 & -1 \\ -1 & 0 \end{bmatrix}\begin{bmatrix} x_1 \\ x_2 \end{bmatrix} + \begin{bmatrix} 1 & 0 \\ 0 & -1 \end{bmatrix}\begin{bmatrix} u_1 \\ u_2 \end{bmatrix}$$

(8)

By using the state-space averaging method [20], the average matrix A and B are calculated as given by Equations (9) and (10).

$$A = A_1 D_r + A_2(1 - D_r),$$

(9)

$$B = B_1 D_r + B_2(1 - D_r),\tag{10}$$

where D_r is the duty cycle of the boost converter.

The steady-state X is calculated as follows:

$$X = A^{-1}BU = \left[\begin{array}{cc} \frac{I_{MXr}}{1-D_r} & V_{inr} \end{array}\right]^{-1}.\tag{11}$$

The small-signal equation becomes as given by Equation (12):

$$\hat{X} = A\hat{x} + B\hat{u} + [(A_1 - A_2)X + (B_1 - B_2)U]\hat{d_r},\tag{12}$$

where $A_1 - A_2 = \left[\begin{array}{cc} 0 & 1 \\ -1 & 0 \end{array}\right]$, $B_1 - B_2 = 0$, and $\hat{d_r}$ is the small signal variation of duty cycle around its steady state operation point.

Equation (12) can be rewritten as Equation (13):

$$\begin{cases} L_{MXr}\frac{d\hat{i}_{inr}}{dt} = -(1 - D_r)v_{MXr} + \hat{v}_{inr} + \frac{V_{inr}}{1-D_r}\hat{d_r} \\ C_{MXr}\frac{d\hat{V}_{MXr}}{dt} = (1 - D_r)\hat{i}_{inr} - \hat{i}_{MXr} - \frac{I_{MXr}}{1-D_r}\hat{d_r} \end{cases}.\tag{13}$$

To simply the analysis, the AC small-signal variation of $\hat{v}_{inN}$ and $\hat{i}_{MXN}$ is assumed to be 0 (negligible) because the dynamic variation of battery voltage and battery module output current is very slow compared to the dynamic variation of the control signal $\hat{d_r}$ (duty cycle) of the power converter. Therefore, by performing the Laplace transformation, Equation (14) can be derived as:

$$\begin{cases} sL_{MXr}i_{inr}(s) = -(1 - D_r)v_{MXr}(s) + \frac{V_{inr}}{1-D_r}d_r(s) \\ sC_{MXr}v_{MXr}(s) = (1 - D_r)i_{inr}(s) - \frac{I_{MXN}}{1-D_r}d_r(s) \end{cases}.\tag{14}$$

Based on Equation (15), the output voltage to the control signal transfer function of the power converter can be derived as

$$G_{vdr} = \frac{v_{MXr}(s)}{d_r(s)} = \frac{\frac{1}{C_{MXr}}\left(-\frac{I_{MXr}}{1-D_r}s + \frac{V_{inr}}{C_{MXr}}\right)}{s^2 + \frac{(1-D_r)^2}{L_{MXr}C_{MXr}}}.\tag{15}$$

Similarly, the input current to control signal transfer function can be derived as

$$G_{idr} = \frac{i_{inr}(s)}{d_r(s)} = \frac{\frac{1}{L_{MXr}}\left(\frac{I_{MXr}}{C_{MXr}} + \frac{V_{inr}}{1-D_r}s\right)}{s^2 + \frac{(1-D_r)^2}{L_{MXr}C_{MXr}}}.\tag{16}$$

2.3.2. Transfer Function of WPT Stage (Half-Bridge Inverter, WPT Coils, and Half-Bridge Rectifier)

Figure 5 shows the circuit diagram of the WPT stage. By writing the Kirchhoff's voltage law (KVL) equations as given by Equation (17), the ratio between the output voltage V_{Rr} and the input voltage V_{Tr} at the resonance frequency can be calculated as given by Equation (18).

$$\left[\begin{array}{cc} j\omega L_{Tr} + \frac{1}{j\omega C_{Tr}} + R_{pTr} & -j\omega M_{TR} \\ -j\omega M_{TR} & j\omega L_{Rr} + \frac{1}{j\omega C_{Rr}} + R_{pTr} + R_{Lr} \end{array}\right]\left[\begin{array}{c} I_{Tr} \\ I_{Rr} \end{array}\right] = \left[\begin{array}{c} V_{Tr} \\ 0 \end{array}\right],\tag{17}$$

$$G_{vTRr}|_{\omega=\omega_o} = \frac{V_{Rr}}{V_{Tr}} = \frac{j\omega M_{TRr}R_{Lr}}{Z_{Tr}Z_{Rr} + (\omega M_{TR})^2} = \frac{j\omega k_{TRr}\sqrt{L_{Tr}L_{Rr}}R_{Lr}}{Z_{Tr}Z_{Rr} + (\omega k_{TRr})^2 L_{Tr}L_{Rr}},\tag{18}$$

where $Z_{Tr} = j\omega L_{Tr} + \frac{1}{j\omega C_{Tr}} + R_{pTr}$ is the equivalent impedance of Tx side and $Z_{Rr} = j\omega L_{Rr} + \frac{1}{j\omega C_{Rr}} + R_{pTr} + R_{Lr}$ is the equivalent impedance of Tx side. By substituting $s = j\omega$ into Equation (19), the following transfer function can be obtained.

$$G_{vTRr}(s) = \frac{V_{Rr}(s)}{V_{Tr}(s)} = -\frac{sk_{TRr}\sqrt{L_{Tr}L_{Rr}}R_{Lr}}{Z_{Tr}Z_{Rr} + s^2 k_{TRr}{}^2 L_{Tr}L_{Rr}}. \tag{19}$$

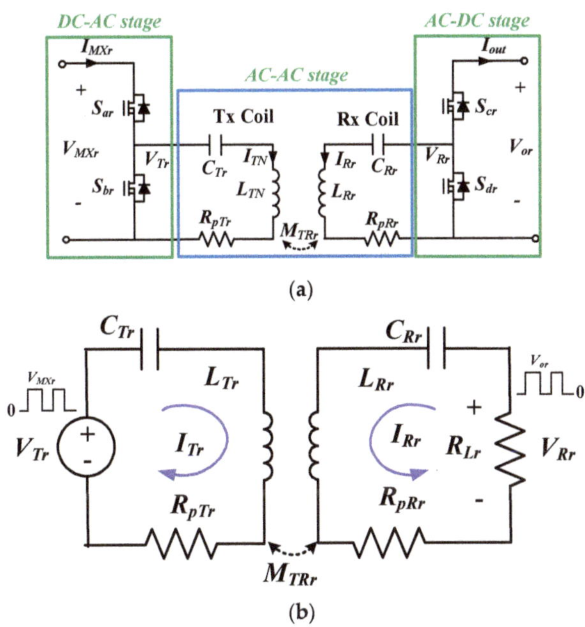

Figure 5. Circuit diagram of the wireless power transfer (WPT) stage in the WEDES system (**a**) equivalent circuit and (**b**) T-model.

Similarly, the ratio between the Tx current and Rx current can be derived as given by Equation (20).

$$G_{iTRr}(s) = \frac{I_{Rr}(s)}{I_{Tr}(s)} = -\frac{s^2 k_{TRr}C_{Rr}\sqrt{L_{Tr}L_{Rr}}}{s^2 L_{Rr}C_{Rr} + sC_{Rr}(R_{pTr} + R_{Lr}) + 1}. \tag{20}$$

3. Compensator Design

3.1. Battery Module Voltage Control Loop Compensator Design

The WEDES system design parameters are shown in Table 1. Based on the small-signal model in Figure 3, the uncompensated discrete-time transfer function of the battery module voltage loop for the rth battery module G_{busr} (Z) consists of G_{PWM}, G_{vdr}, G_{vTRr}, k_{divr}, ZOH_{vr}, and digital computation delay $Delay_{dr}$. G_{busN} (Z) is calculated as given by Equation (21), and its bode plot is shown as the dashed curve in Figure 6.

$$L_{MXr_{uncomp}}(z) = Z\{G_{PWM}(s){\cdot}G_{vdr}(s){\cdot}G_{vTRr}(s){\cdot}ZOH_{vN}(s){\cdot}Delay_{dN}\} = \frac{-0.0001999z - 0.000272}{z^3 - 1.977z^2 + 0.9789z}, \tag{21}$$

where $ZOH_{vr}(s) = \frac{1-e^{-s \cdot T_s}}{s}$; $Delay_{dr}(s) = e^{-sT_{delay}}$; T_{delay} is the digital controller computation delay and it is equal to 10 μs in the experimental implementation; $G_{PWM} = 1/1024$; $K_{div} = 11$ with 1 kΩ and 10 kΩ resistors as the voltage divider.

Table 1. Design parameters of the wireless distributed and enabled battery energy storage (WEDES) system.

Parameter	Value	Parameter	Value
V_{MXr}	0–20 V	L_T, L_R	24 μH
$V_{bus\text{-}ref}$	30 V	C_T, C_R	0.4 μF
I_{out}	2 A	k_{TR}	0.3
L_{MXr}	47 μH	R_{pT}, R_{RT} @ 50 kHz	0.1 Ω
C_{MXr}	220 μF	f_{sw_bridge}	50 kHz
$f_{sw\_converter}$	100 kHz	-	-

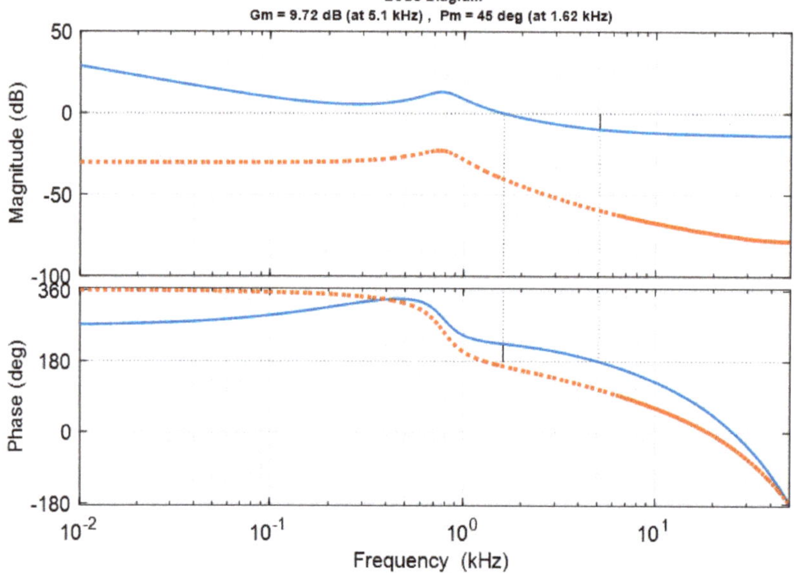

Figure 6. The bode plot of the uncompensated (red-dashed curve) and compensated (blue-solid curve) battery module voltage control loop gain.

In the battery module voltage compensator design, there is a right-hand-plane-zero (RHPZ), which is located at 4.86 kHz. This RHPZ is introduced due to the existing boost converter. To guarantee the stability of the system, the compensated control bandwidth should be smaller than RHPZ. With a compensator $G_{v\text{-}bus}(z)$ given by (22), the compensated battery module output voltage control loop gain ($G_{MXr\_comp} = G_{MXr\_uncomp}(z) \cdot G_{MXr}(z)$) achieves a control bandwidth of 1.62 kHz and a phase margin of 45°, as shown on the solid curve in Figure 6.

$$G_{MXr}(z) = \frac{950.8z^2 - 1850z + 899.7}{z^2 - z} \tag{22}$$

3.2. SOC Balancing Loop Compensator Design

According to the small-signal model shown in Figure 3, the uncompensated SOC loop gain (i.e., with unity SOC loop compensator gain) is given by Equation (23), and its bode plot is represented by the dashed curve in Figure 7.

$$L_{socr-uncomp}(z) = Z\{G_{idN}(s){\cdot}ZOH_{icellN}(s){\cdot}G_{sociN}(s){\cdot}ZOH_{socN}(s){\cdot}G_{socN}{\cdot}(-1){\cdot}\delta_{MXN}{\cdot}G_{\alpha\lambda}{\cdot}Delay_w{\cdot}G_{v-bus}\}, \qquad (23)$$

where $ZOH_{icellN}(s) = \frac{1-e^{-s{\cdot}T_s}}{s}$, $ZOH_{socN}(s) = \frac{1-e^{-s{\cdot}T_{soc}}}{s}$, T_{soc} is the sampling period for the SOC value in the SOC balancing loop. Since the SOC value of a battery cell varies very slowly compared to the switching period of the power converter, the sampling rate of the SOC balancing loop does not have to be very fast. T_{soc} =1 s is found to be a good trade-off between the hardware resource consumption, system stability and SOC balancing speed.

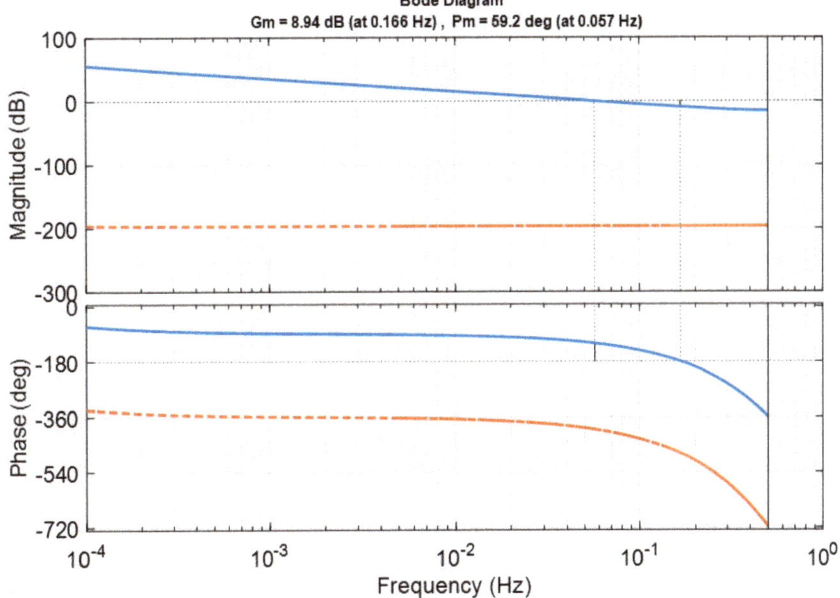

Figure 7. The bode plot of uncompensated (red-dashed curve) and compensated (blue-solid curve) SOC balancing control loop.

With a compensator given by Equation (24), the compensated SOC balancing loop gain achieves a control bandwidth of 0.057 Hz and phase margin of 59.2°, as shown on the solid curve in Figure 7. Due to the slow sampling rate of SOC value (1 Hz), it is expected that the control bandwidth of the SOC balancing loop is much lower than that of the voltage control loop.

$$G_{SOCr}(z) = \frac{2.5\times10^8}{z-1} \qquad (24)$$

3.3. Bus Voltage Control Loop Compensator Design

Based on the small-signal model shown in Figure 3, the uncompensated bus voltage control loop gain (i.e., with gain = 1) is given by Equation (25), and its bode plot is represented by the dashed curve in Figure 8. With a compensator given by Equation (26), the compensated bus voltage loop gain achieves a control bandwidth of 8.12 kHz and a phase margin of 52.3°, as shown on the solid curve in Figure 8.

$$L_{bus-uncomp}(z) = Z\{ZOH_{v-bus}(s)\cdot Delay_w\} = \frac{0.2555z - 0.1798}{z^2 - 1.977z + 0.9789}, \tag{25}$$

$$G_{bus}(z) = \frac{1.2z - 1.8}{z - 1}, \tag{26}$$

where $ZOH_{v-bus}(s) = \frac{1-e^{-s\cdot T_s}}{s}$.

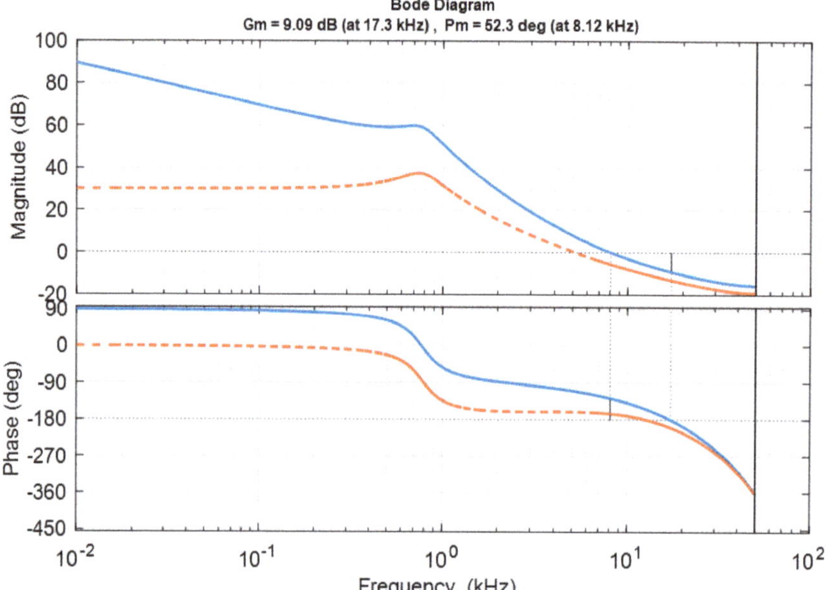

Figure 8. The bode plot of the uncompensated (red-dashed curve) and compensated (blue-solid curve) bus voltage control loop gain.

4. Simulation and Model Experiment Validation

In this section, the small-signal model was evaluated and validated in both simulation and hardware experiments. The simulation model was built in MATLAB®/SIMULINK software (2018a, MatchWorks, Natick, Massachusetts, USA) using the derived transfer functions in Sections 2 and 3. The hardware control compensator was implemented with Texas Instrumental microcontroller TMS320S28335. The design parameters are shown in Table 1. In the hardware experiment, three WEDES battery modules were implemented and utilized.

For the verification of derived small-signal models of the WEDES system, the dynamic response of both simulation and hardware experiments under different operation conditions was compared. If the dynamic performance from simulation results and experimental results are in good agreement, it can be implied that the developed model is valid.

As discussed in previous sections, there are three different control loops (battery module voltage control loop, bus voltage control loop, and SOC balancing control loop) in the WEDES system. During the test for a specific control loop, one of the operation parameters was changed while the rest of the operation parameters were set constant.

4.1. Experimental Results for Battery Module Voltage Control Loop

For the test of battery module voltage control loop, the reference output voltage for each battery module was changed suddenly from 10 V ($V_{MX1\text{-}DC\text{-}ref}$ = $V_{MX2\text{-}DC\text{-}ref}$ = $V_{MX3\text{-}DC\text{-}ref}$ = 10 V under steady-state operation) to $V_{MX1\text{-}DC\text{-}ref}$ = 13 V, $V_{MX2\text{-}DC\text{-}ref}$ = 9 V and $V_{MX3\text{-}DC\text{-}ref}$ = 8 V. The simulation results and experimental results are shown in Figure 9.

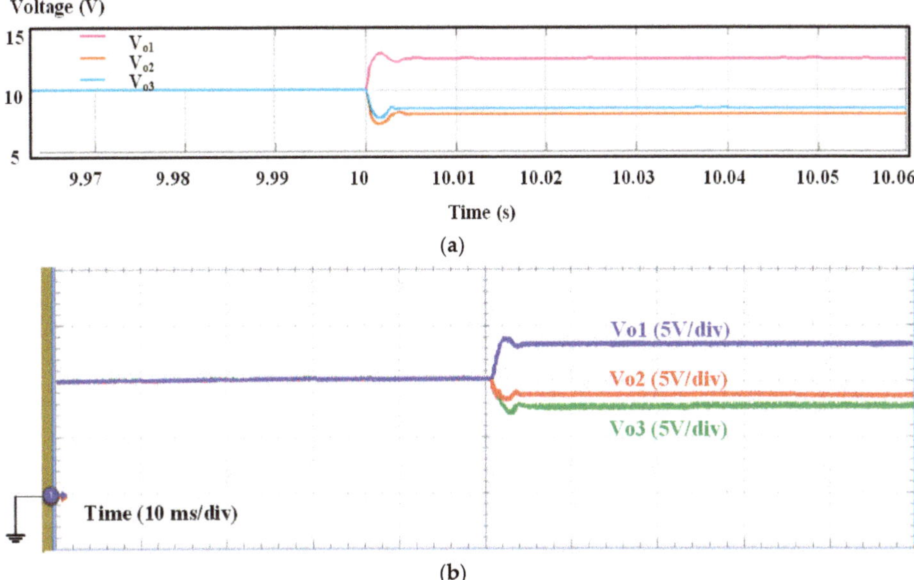

Figure 9. Waveforms for three battery module output voltage when the reference voltages were changed from $V_{MX1\text{-}DC\text{-}ref}$ = $V_{MX2\text{-}DC\text{-}ref}$ = $V_{MX3\text{-}DC\text{-}ref}$ = 10 V to $V_{MX1\text{-}DC\text{-}ref}$ = 13 V, $V_{MX2\text{-}DC\text{-}ref}$ = 9 V and $V_{MX3\text{-}DC\text{-}ref}$ = 8 V, (**a**) simulation results and (**b**) hardware experimental results.

From Figure 9, it can be observed that once the reference voltage values were changed, the output voltage of battery module 1 was controlled to increase while the output voltages of battery modules 2 and 3 were controlled to decrease. In Figure 9, the simulation model results and the hardware experimental results are in good agreement. In other words, the shape, magnitude, and overshoot/undershoot of the waveforms of simulation and hardware experiments match each other, which validates the small-signal model for the battery module voltage loop.

4.2. Experimental Results for SOC Balancing Control Loop

For the test of SOC balancing control loop, the SOC value of three battery modules were changed suddenly from SOC_1 = SOC_2 = SOC_3 = 70% under balanced conditions to SOC_1 = 75%, SOC_2 = 70%, and SOC_3 = 65%. The experimental results are shown in Figure 10.

It can be observed from Figure 10 that once there was a change in SOC values, V_{o1} increased while V_{o3} decreased. This is because the value of SOC_1 was larger than the average SOC value of three battery modules (i.e., (75% + 70% + 65%)/3 = 70%), therefore resulted in a larger value of the battery module output voltage V_{o1} for faster discharge. The variation of V_{o3} for battery module 3 followed the same behavior but in the opposite direction. The value of SOC_2 was equal to the average SOC value, and therefore, its corresponding battery module output voltage remained constant. Figure 10 also shows that the simulation results and hardware experimental results are in good agreement under the variation of SOC values test condition.

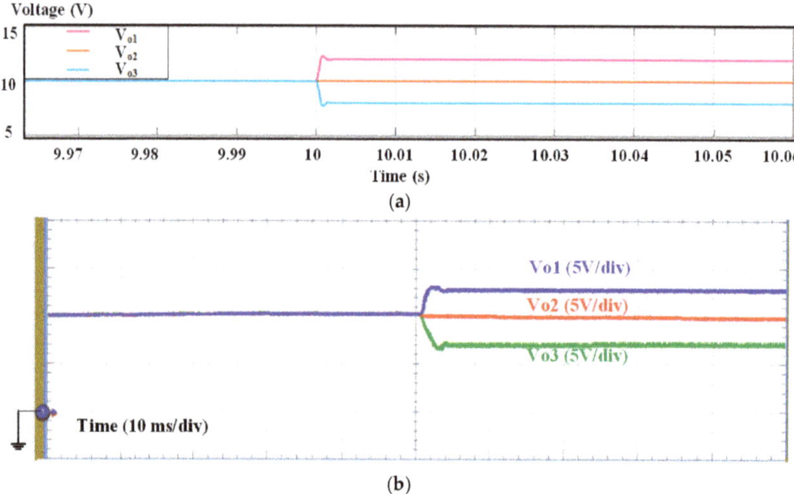

Figure 10. Waveforms for three battery module output voltage when SOC values of three modules changed from $SOC_1 = SOC_2 = SOC_3 = 70\%$ to $SOC_1 = 75\%$, $SOC_2 = 70\%$, and $SOC_3 = 65\%$, (**a**) simulation results and (**b**) hardware experimental results.

4.3. Experimental Results for Bus Voltage Control Loop

The last test was implemented for the bus voltage control loop. In this test, the bus voltage reference value was changed suddenly from 30 V to 37.5 V, as shown in Figure 11. It can be observed from Figure 11 that as the bus voltage reference value increased, the battery module voltage values changed from 10 V to 12.5 V correspondingly, which yielded a total bus voltage equal to its new reference value. The waveforms shown in Figure 11 demonstrate the consistency between simulation results and hardware experimental results.

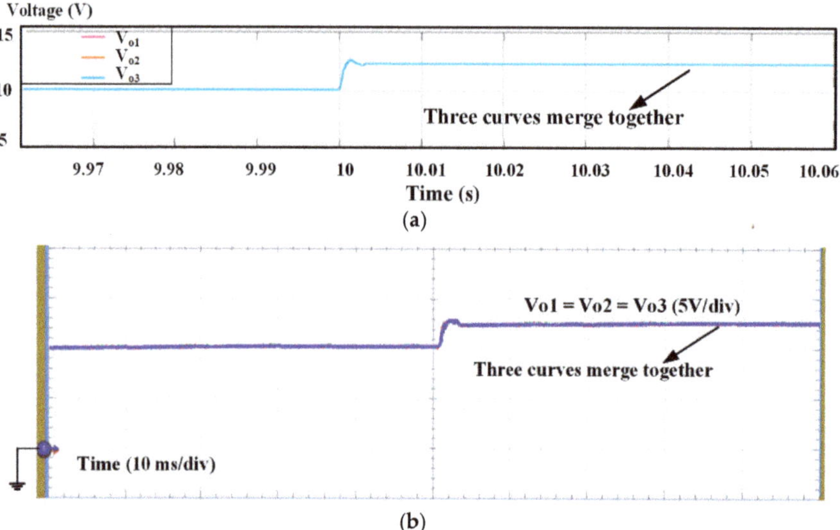

Figure 11. Waveforms for three battery module output voltage when bus reference voltage changes from $V_{bus\text{-}ref} = 30$ V to $V_{bus\text{-}ref} = 37.5$ V, (**a**) simulation results and (**b**) hardware experimental results.

5. Additional Comment

It should be noted that the presented WEDES system in this paper is different from a traditional Inductor-Capacitor (LC) compensation wireless power transfer (WPT) system or a traditional battery energy storage system (BESS). Instead, it is a combination of both systems. Table 2 shows a comparison between the presented WEDES system in this paper, conventional WPT system [18,21,22], and conventional BESS system [23–25]. It is shown that the presented WEDES system combines the advantages of both the conventional WPT system and BESS system, and can achieve bus voltage regulation and SOC balancing, and allow for fast and safe exchange/swapping of smaller and lighter battery modules with WPT technology.

Table 2. Comparison summary between the WEDES system, conventional Inductor-Capacitor (LC) compensation wireless power transfer (WPT) system, and conventional battery energy storage system (BESS).

Parameter	WEDES System in this Paper	Conventional LC Compensation WPT System [18,21,22]	Conventional BESS System [23–25]
Typical Components/Devices	Battery cells, power converter, inverter, WPT coils, and rectifier	Battery cells, inverter, WPT coils, and rectifier	Battery cells and power converter/inverter
System Complexity	Medium	Medium	Low
Able to Realize SOC Balancing	Yes	No	Yes
Able to Realize Voltage Regulation	Yes	Yes	Yes
Typical Battery Operation	Discharging	Charging	Discharging
Able to Transfer Power Wirelessly	Yes	Yes	No
Able to Insert/Remove Battery Module during Normal Operation	Yes	No	No
Efficiency	High (>85%)	High (>90% at short distance)	Highest (>95%)

6. Conclusions

In this paper, the small-signal modeling of a distributed WEDES battery system was derived to analyze the steady-state stability and dynamic response of the entire system, as well as provide guidelines for the controller design of multiple interacted control loops. Based on the small-signal models and associated transfer functions, all three control loops, including battery module output voltage control loop, SOC balancing control loop, and bus voltage control loop with compensators, were evaluated and validated by both simulations and a 3-module WEDES battery system. It was shown that the experimental results from simulation and hardware prototype were in good agreement, which validates the accuracy and effectiveness of the derived small-signal model and designed compensators.

Funding: This research received no external funding.

Conflicts of Interest: The author declares no conflict of interest.

References

1.   Abu Qahouq, J.; Cao, Y. Control scheme and power electronics architecture for a wirelessly distributed and enabled battery energy storage system. *Energies* **2018**, *11*, 1887. [CrossRef]

2. Lelie, M.; Braun, T.; Knips, M.; Nordmann, H.; Ringbeck, F.; Zappen, H.; Sauer, D.U. Battery management system hardware concepts: An overview. *Appl. Sci.* **2018**, *8*, 534. [CrossRef]

3. Charging Station (Battery Swapping section). Wikipedia, the Free Encyclopedia. Available online: https://en.wikipedia.org/wiki/Charging_station#Battery_swapping (accessed on 24 August 2019).

4. Shao, S.; Guo, S.; Qiu, X. A mobile battery swapping service for electric vehicles based on a battery swapping van. *Energies* **2017**, *10*, 1667. [CrossRef]

5. Hou, J.; Yang, Y.; He, H.; Gao, T. Adaptive Dual Extended Kalman Filter Based on Variational Bayesian Approximation for Joint Estimation of Lithium-Ion Battery State of Charge and Model Parameters. *Appl. Sci.* **2019**, *9*, 1726. [CrossRef]

6. Cao, Y.; Abu Qahouq, J.A. Analysis and Evaluation of a Dual-Variable Closed-Loop Control of Power Converter with Wireless and Nonwireless Power Transfer. *IEEE Trans. Ind. Electron.* **2019**, *66*, 2668–2679. [CrossRef]

7. Chen, J.; Chen, C.; Duan, S. Cooperative Optimization of Electric Vehicles and Renewable Energy Resources in a Regional Multi-Microgrid System. *Appl. Sci.* **2019**, *9*, 2267. [CrossRef]

8. Rakhymbay, A.; Khamitov, A.; Bagheri, M.; Alimkhanuly, B.; Lu, M.; Phung, T. Precise analysis on mutual inductance variation in dynamic wireless charging of electric vehicle. *Energies* **2018**, *11*, 624. [CrossRef]

9. Schroeder, A.; Traber, T. The economics of fast charging infrastructure for electric vehicles. *Energy Policy* **2012**, *43*, 136–144. [CrossRef]

10. Kong, P.Y.; Karagiannidis, G. Charging schemes for plug-in hybrid electric vehicles in smart grid: A. survey. *IEEE Access* **2016**, *4*, 6846–6875. [CrossRef]

11. Deutsches Elektronen-Synchrotron DESY. Fast Charging Cycles Make Batteries Age More Quickly: X-Ray Study Images Damage in Lithium-Ion Batteries. Science Daily. April 2015. Available online: www.sciencedaily.com/releases/2015/04/150414094117.htm (accessed on 24 August 2019).

12. Chandrasekaran, R. Quantification of bottlenecks to fast charging of lithium-ion-insertion cells for electric vehicles. *J. Power Sources* **2014**, *271*, 622–632. [CrossRef]

13. Whitney, L. Tesla Demos Electric Battery Swap in just 90 Seconds. CNET. June 2013. Available online: https://www.cnet.com/news/tesla-demos-electric-battery-swap-in-just-90-seconds/ (accessed on 24 August 2019).

14. Miller, J.M.; Jones, P.T.; Li, J.M.; Onar, O.C. ORNL experience and challenges facing dynamic wireless power charging of EV's. *IEEE Circuits Syst. Mag.* **2015**, *15*, 40–53. [CrossRef]

15. Zhu, Q.; Wang, L.; Liao, C. Compensate capacitor optimization for kilowatt-level magnetically resonant wireless charging system. *IEEE Trans. Ind. Electron.* **2014**, *61*, 6758–6768. [CrossRef]

16. Zhang, L.; Peng, H.; Ning, Z.; Mu, Z.; Sun, C. Comparative research on RC equivalent circuit models for lithium-ion batteries of electric vehicles. *Appl. Sci.* **2017**, *7*, 1002. [CrossRef]

17. Luo, Z.; Wei, X. Analysis of square and circular planar spiral coils in wireless power transfer system for electric vehicles. *IEEE Trans. Ind. Electron.* **2018**, *65*, 331–341. [CrossRef]

18. Cao, Y.; Abu Qahouq, J.A. Modelling and control design of reconfigurable wireless power transfer system for transmission efficiency maximization and output voltage regulation. *IET Power Electron.* **2019**, *12*, 1903–1916. [CrossRef]

19. Dang, Z.; Cao, Y.; Abu Qahouq, J.A. Reconfigurable Magnetic Resonance-Coupled Wireless Power Transfer System. *IEEE Trans. Power Electron.* **2015**, *30*, 6057–6069. [CrossRef]

20. Erickson, R.W.; Maksimovic, D. *Fundamentals of Power Electronics*; Springer: Berlin/Heidelberg, Germany, 2001; pp. 265–330.

21. Wang, Y.; Lin, F.; Yang, Z.P.; Liu, Z.Y. Analysis of the Influence of Compensation Capacitance Errors of a Wireless Power Transfer System with SS Topology. *Energies* **2017**, *10*, 2177. [CrossRef]

22. Galigekere, V.P.; Pries, J.; Onar, O.C.; Su, G.; Anwar, S.; Wiles, R.; Seiber, L.; Wilkins, J. Design implementation of an optimized 100 kW stationary wireless charging system for EV battery recharging. In Proceedings of the 2018 IEEE Energy Conversion Congress and Exposition (ECCE), Portland, OR, USA, 23–27 September 2018; pp. 3587–3592.

23. Cao, Y.; Abu Qahouq, J.A. Evaluation of bi-directional single-inductor multi-input battery system with state-of-charge balancing control. *IET Power Electron.* **2018**, *11*, 2140–2150. [CrossRef]

24. Efstratios, C.; Rogers, D.J. Cell SoC balancing using a cascaded full bridge multilevel converter in battery energy storage systems. *IEEE Trans. Ind. Electron.* **2016**, *63*, 5394–5402.
25. Qian, H.; Zhang, J.; Lai, J.S.; Yu, W. A high-efficiency grid-tie battery energy storage system. *IEEE Trans. Power Electron.* **2011**, *26*, 886–896. [CrossRef]

Article

SOC Estimation with an Adaptive Unscented Kalman Filter Based on Model Parameter Optimization

Xiangwei Guo [1],*, Xiaozhuo Xu [1],*, Jiahao Geng [1], Xian Hua [2], Yan Gao [1] and Zhen Liu [1]

[1]   School of Electrical Engineering and Automation, Henan Polytechnic University, Jiaozuo 454003, China; gjh8094@126.com (J.G.); hpugy@foxmail.com (Y.G.); jxx@vip.163.com (Z.L.)
[2]   The Battery Research Institute of Henan Province, Xinxiang 453000, China; huaxianhpu@outlook.com
*   Correspondence: gxw@hpu.edu.cn (X.G.); xxzhpu@163.com (X.X.); Tel.: +86-186-0391-2411 (X.G.)

Received: 27 August 2019; Accepted: 2 October 2019; Published: 6 October 2019

Featured Application: **The method of the research can be applied to the estimation of the remaining battery charge of electric vehicles.**

Abstract: State of charge (SOC) estimation is generally acknowledged to be one of the most important functions of the battery management system (BMS) and is thus widely studied in academia and industry. Based on an accurate SOC estimation, the BMS can optimize energy efficiency and protect the battery from being over-charged or over-discharged. The accurate online estimation of the SOC is studied in this paper. First, it is proved that the second-order resistance capacitance (RC) model is the most suitable equivalent circuit model compared with the Thevenin and multi-order models. The second-order RC equivalent circuit model is established, and the model parameters are identified. Second, the reasonable optimization of model parameters is studied, and a reasonable optimization method is proposed to improve the accuracy of SOC estimation. Finally, the SOC is estimated online based on the adaptive unscented Kalman filter (AUKF) with optimized model parameters, and the results are compared with the results of an estimation based on pre-optimization model parameters. Simulation experiments show that, without affecting the convergence of the initial error of the AUKF, the model after parameter optimization has a higher online SOC estimation accuracy.

Keywords: SOC; second-order RC model; model parameter optimization; AUKF

---

1. Introduction

The power battery State of charge (SOC) is generally defined as the ratio between the available capacity and the reference capacity. An accurate SOC estimation is an important basis for the formulation of an optimal energy management strategy for the whole vehicle control system of an electric vehicle (EV). It is of great significance for extending the life of the battery pack and improving the safety of the battery system [1,2]. Due to the nonlinear nature of the power battery, the SOC cannot be directly acquired by sensors; rather, it must be estimated by measuring physical quantities, such as the battery voltage, operating current, and internal resistance of the battery, and by using certain mathematical methods [3,4].

Commonly used estimation methods include the open-circuit voltage method, the ampere-hour integration method, the neural network method and the Kalman filter method [5–9]. The open-circuit voltage (OCV) can represent the discharge capacity of a lithium battery in its current state. It has a good linear relationship with the SOC when the SOC is greater than 0.1. The OCV cannot be directly measured during the working state of the battery; it can only be approximated when the battery is not working. Therefore, this method is only applicable to EVs that are parked, and the OCV method is used to provide an initial value of the SOC for other estimation methods. The ampere-hour (Ah) integral method is a relatively common, simple and reliable method for estimating the SOC. When the discharge current is positive and the charging current is negative, the calculation formula can be expressed as:

$$SOC_t = SOC_0 - \frac{1}{C_N} \int_0^t \eta i dt \tag{1}$$

where SOC_0 is the initial SOC value, C_N is the maximum available capacity of the battery, which is almost invariant if the time scale is small and battery ageing is accordingly ignored [8], and η is the coulomb efficiency. The ampere-hour integral method has the advantages of low cost and convenient measurement. However, there are several problems in the application of this method in EVs: (1) Other methods are needed to obtain the initial value of the SOC. (2) The accuracy of the current measurement has a decisive influence on the accuracy of the SOC estimation. (3) The cumulative error of the integration process cannot be eliminated, and if the charging or discharging time is too long during one calculation, cumulative errors can cause estimates to be unreliable.

A neural network (NN) is an intelligent mathematical tool [10,11]. A NN has the adaptability and self-learning skills to demonstrate a complex nonlinear model. NNs use trained data to estimate the SOC without knowing any information about the internal structure of the battery and the initial SOC. Theoretically, the nonlinear characteristics of the power battery can be better mapped by a NN. The advantage of this method is that it is capable of working in nonlinear battery conditions while the battery is charging/discharging. Nevertheless, the algorithm needs to store a large amount of data for training, which not only requires large memory storage but also overloads the entire system.

The Kalman filter method [12–14] is the current research hotspot for SOC estimation. The core idea of the Kalman filter is the optimal estimate in the sense of minimum variance, including the two stages of prediction and updating. In the prediction stage, the filter applies the value of the previous state to estimate the current state, and in the updating stage, the filter optimizes the predicted value in the prediction stage by using the observed value in the current state to obtain a more accurate estimate of the current state. It should be noted that the basic Kalman filter is mainly applied to linear systems. However, the estimation of the SOC is related to many factors, such as charge and discharge current, and cut-off voltage, and the influence of these factors on the SOC is nonlinear. For this reason, some people have used the extended Kalman filter (EKF) to estimate the battery SOC. Although some achievements have been made, there is a linearization error in actual use, and the Jacobian matrix is difficult to estimate. In recent years, a novel derivative Kalman filter algorithm, the unscented Kalman filter (UKF), has emerged to realize nonlinear filtering. The UKF directly uses nonlinear unscented transform techniques without the need for Taylor approximations of nonlinear equations. This process allows the mean and variance of the state of the nonlinear system to propagate directly according to the nonlinear mapping, thereby achieving a higher estimation accuracy.

In the traditional UKF algorithm [15–19], the covariance is a constant and cannot satisfy the real-time dynamic characteristics of the noise, which has a certain impact on the accuracy [20,21]. To eliminate this effect, the traditional UKF algorithm is improved by updating the covariance in real-time, which thus improves the accuracy of the UKF in this paper. This type of algorithm is called the adaptive unscented Kalman filter (AUKF) algorithm. On this basis, a reasonable optimization of the model parameters is studied, and a reasonable optimization method is proposed to further improve the accuracy of SOC estimation.

The structure of this paper is arranged as follows. In Section 1, the most commonly used methods for SOC estimation are introduced, and the proposed method of this paper is briefly described. In Section 2, the battery model is established, the model parameters are identified, and the model accuracy is verified. In Section 3, the AUKF based on a second-order resistance capacitance (RC) equivalent circuit model is presented. In Section 4, the basis and method of the reasonable optimization of the model parameters are put forward. In Section 5, the accuracy and initial error convergence of the AUKF before and after model parameter optimization are compared by simulation experiments, and the advantages of the proposed method are demonstrated. Finally, in Section 6, the research results of this paper are summarized, and future research directions are provided.

2. Establishment of Battery Model and Parameter Identification

The open-circuit voltage and internal resistance are the most basic components of an equivalent circuit model. All equivalent circuit models add other components on the basis of these two components to improve the model accuracy. Typical equivalent circuit models of lithium-ion batteries include the Rint model, the Thevenin model, the PNGV (Partnership for a New Generation of Vehicles) model and the multi-order RC loop model [22–25]. For the complicated polarization characteristics of a battery, some studies deduce and suggest that the models with more parallel RC networks connected in series should have a much higher accuracy; however, at the same time, memory is consumed, waste calculations are performed and bad real-time applications are caused.

2.1. Establishment of Lithium-Ion Battery Model

The measured terminal voltage response curve of the Sanyo lithium battery with a rated capacity of 2.6 AH at a certain state (SOC = 0.9 with a constant current discharge of 260 mAh at 0.5 C) at the end of discharge is shown in Figure 1.

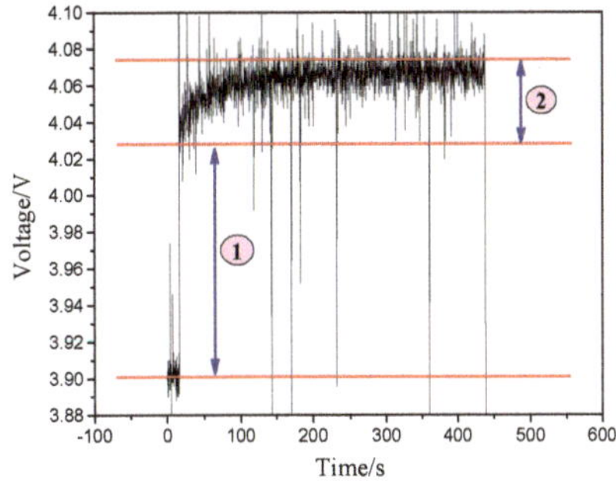

Figure 1. Terminal voltage response at the end of discharge.

The polarization effect of a lithium battery is usually expressed equivalently by an RC loop and is equivalent to the first-order zero input response after the battery is discharged. Its terminal voltage can be expressed as an exponential term ($IRe^{\frac{-t}{\tau}}$). Region 2 in Figure 1 is the disappearance process of the polarization effect. Exponential curves are used to fit the single index, double index and triple index coefficients of region 2, and the fitting results are shown in Figure 2.

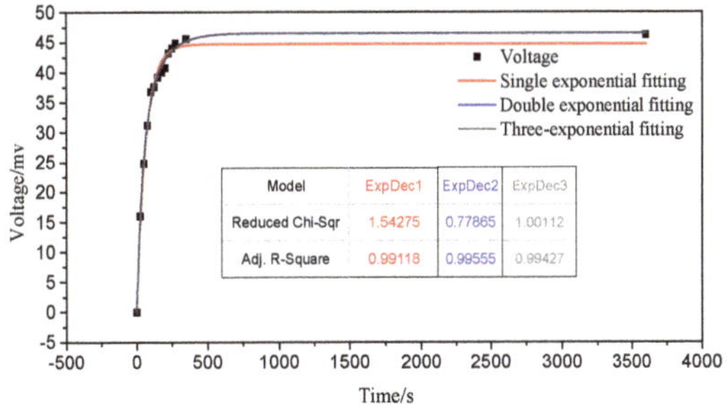

Figure 2. Coefficient fitting of exponential curves.

It can be seen from Figure 2 that the reduced chi-squared statistics of the double and triple exponential fittings are smaller than that of the single exponential fitting. That is, the random error effect of the double and triple exponential fittings are smaller than that of the single exponential fitting. At the same time, the correction decision coefficient of the double and triple exponential fittings is larger than that of the single exponential fitting and is closer to 1, and the fitting effect is better. Therefore, the double and triple exponential fittings can more accurately reflect the polarization effect of the battery. By comparing the results of the double and triple exponential fittings, the sum of the squares of the residuals of the double exponential fitting is less than that of the triple exponential fitting, and the correction decision coefficient of the double exponential fitting is also closer to 1 than that of the triple exponential fitting. The results show that the double exponential fitting effect is better than the triple exponential fitting effect. The reason for this fitting result is that although the third-order RC loop can theoretically better reflect the dynamic characteristics of the battery, the third-order RC loop has one more RC loop than the second-order RC loop, which means that the third-order RC loop has two more unknowns than the second-order RC loop in the process of data fitting by computer. Therefore, the fitting effect of the third-order RC loop is not as good as that of the second-order RC loop, and it is proved that the second-order RC model is better than the others for a lithium battery. On this basis, considered comprehensively, the second-order RC equivalent circuit model, as shown in Figure 3, is adopted in this paper.

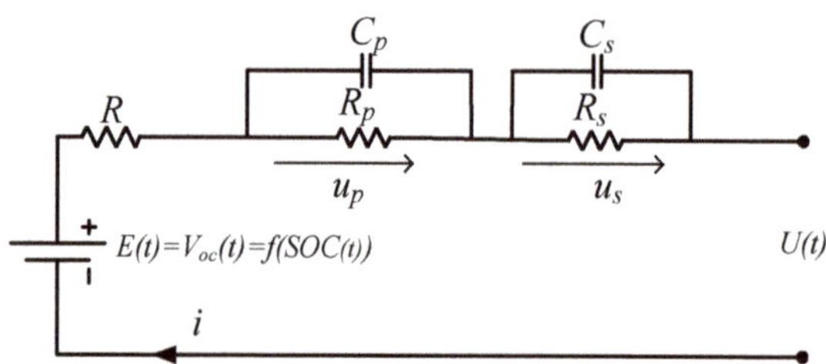

Figure 3. Second-order resistance capacitance (RC)RC battery model.

The model consists of three parts:

Voltage source: V_{oc} represents the open circuit voltage of the power battery. In this paper, the influences of temperature and state of health (SOH) on the OCV are not considered, and the functional relationship between V_{oc} and the battery SOC is studied under the same temperature and SOH conditions. Additionally, the effect of corrosion on the battery is neglected because the focus of this paper is on the equivalent circuit model.

Ohmic resistance: R is used to represent the internal resistance of the battery, which can be determined by the sudden change in voltage after the end of discharge, as shown in region 1 of Figure 1.

RC loop circuit: Two links of a resistor and a capacitor are superposed to simulate battery polarization, which is used to simulate the process of voltage stabilization after discharge. Region 2 of Figure 1 shows the change in voltage influenced by the RC loop circuit.

As shown in Figure 3, the functional relationship of the equivalent circuit model is as follows:

$$\begin{cases} E(t) = iR + u_s + u_p + U(t) = F(SOC(t)) \\ i = \frac{u_s}{R_s} + C_s\frac{du_s}{dt} \\ i = \frac{u_p}{R_p} + C_p\frac{du_p}{dt} \end{cases} \tag{2}$$

For the discretization of Equation (2), the state equation is solved as:

$$\begin{bmatrix} u_{s,k} \\ u_{p,k} \end{bmatrix} = \begin{bmatrix} a_s & 0 \\ 0 & a_p \end{bmatrix}\begin{bmatrix} u_{s,k-1} \\ u_{p,k-1} \end{bmatrix} + \begin{bmatrix} b_s \\ b_p \end{bmatrix}I_{k-1} + \begin{bmatrix} w_3(k) \\ w_5(k) \end{bmatrix} \tag{3}$$

$$U_k = E_k - I_kR - U_{s,k} - U_{p,k} + v(k) = F(SOC_k) - I_kR - U_{s,k} - U_{p,k} + v(k) \tag{4}$$

of which:

$$\begin{cases} a_s = e^{\frac{-T}{R_sC_s}}, \ b_s = R_s - R_se^{\frac{-T}{R_sC_s}} \\ a_p = e^{\frac{-T}{R_pC_p}}, \ b_p = R_p - R_pe^{\frac{-T}{R_pC_p}} \end{cases} \tag{5}$$

2.2. Identification and Verification of Battery Model Parameters

In the process of establishing the battery model, it is necessary to use the approximate linear relationship between the OCV and the SOC. In this section, based on the OCV–SOC relationship curve, the parameters of the second-order RC model are identified, and then the accuracy of the model is verified.

2.2.1. Obtaining the OCV–SOC Relationship Curve

The OCV–SOC curve was obtained from a 18,650 cylindrical lithium-ion battery manufactured by the Sanyo Corporation of Japan with a rated capacity of 2.6 Ah and a rated voltage of 3.7 V, as before. Charge and discharge experiments were carried out on the new battery at a constant temperature. The battery test platform is shown in Figure 4.

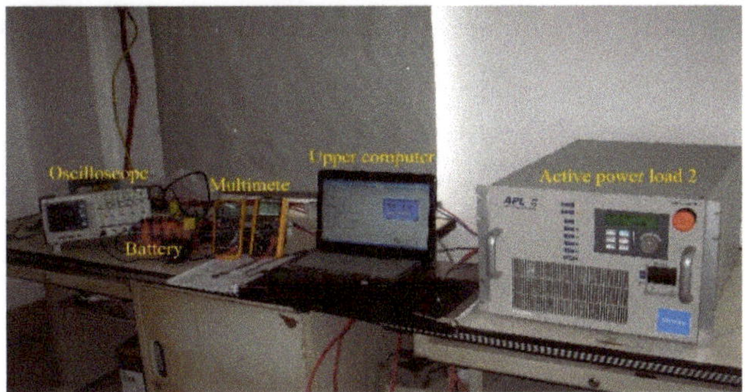

Figure 4. Battery test platform.

The OCV–SOC curves are calibrated under constant current and constant capacity intermittent discharge conditions of 0.2 C, 0.3 C, 0.4 C, 0.5 C, 0.6 C, 0.75 C and 1 C. The specific calibration process of each group is as follows [3]:

1. First, the battery is charged with a constant current (0.2 C), and then the battery is charged with a constant voltage (cut-off voltage of 4.25 V). After charging is completed, the battery is left for one hour to eliminate the polarization effect;
2. The battery is discharged with a constant current and a constant capacity (1/10 of the total capacity, 260 mAh);
3. After the discharge is complete, the battery is left for one hour, and then the open circuit voltage is measured;
4. Repeat steps 2 and 3 until the battery is completely discharged.

The calibration experiment results are shown in Figure 5. When the SOC is greater than 10%, the curves almost coincide. This shows that the OCV–SOC curves corresponding to different discharge rates are approximate under the same temperature (25 °C) and health conditions. In this paper, the OCV–SOC curve under the condition of 0.2 C constant current intermittent discharge is selected as the reference curve.

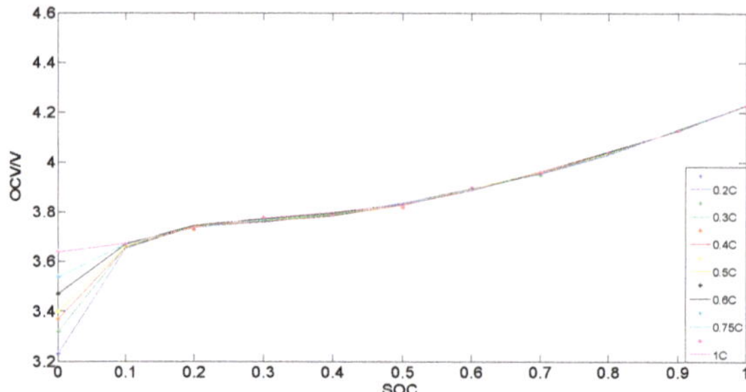

Figure 5. The calibration curve results.

Using MATLAB to fit the polynomial coefficients, the approximate linear relationship between the OCV and the SOC can be obtained, as shown in Equation (6):

$$Voc = a1 \times SOC^6 + a2 \times SOC^5 + a3 \times SOC^4 + a4 \times SOC^3 + a5 \times SOC^2 + a6 \times SOC + a7 \tag{6}$$

where $a_1 = -34.72$, $a_2 = 120.7$, $a_3 = -165.9$, $a_4 = 114.5$, $a_5 = -40.9$, $a_6 = 7.31$, and $a_7 = 3.231$.

2.2.2. Model Parameter Identification

Figure 6 is a schematic diagram of the terminal voltage response curve at the end of lithium battery discharge.

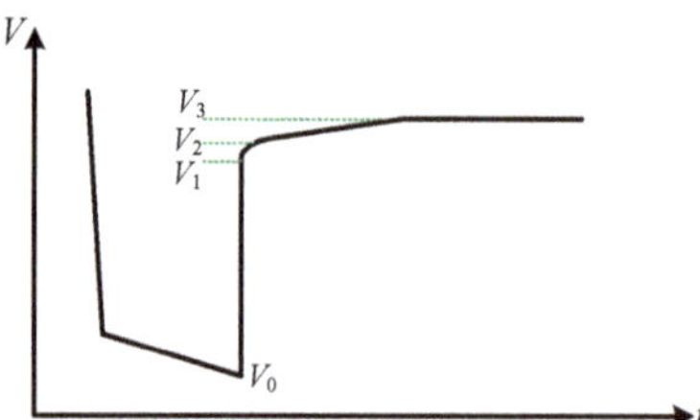

Figure 6. Terminal voltage response curve at the end of battery discharge.

$(V_1 - V_0)$ is the process in which the voltage drop generated on the ohmic resistance inside the battery disappears at the end of discharge. Thus, the ohmic resistance of the battery can be obtained as $R = \frac{V_1 - V_0}{I}$. The battery model adopted in this paper simulates the polarization process of the battery by superposing two RC links. An RC parallel circuit with a small time constant is used to simulate the process of rapid voltage change $(V_2 - V_1)$ when the current of the battery changes abruptly. An RC parallel circuit with a large time constant is used to simulate the process of slow voltage change $(V_3 - V_2)$ when the current of the battery changes abruptly.

The battery is assumed to discharge for a period of time during $(t_0 - t_r)$ and then stay in a resting state for the remaining time. Here, t_0, t_d, and t_r are the discharge start time, the discharge stop time and the static stop time, respectively. In this process, the RC network voltages are:

$$U_s = \begin{cases} R_s i(t)\left[1 - e^{\frac{-(t-t_0)}{\tau_s}}\right] & t_0 < t \leq t_d \\ \\ U_s(t_d)e^{\frac{-(t-t_d)}{\tau_s}} & t_d < t < t_r \end{cases} \tag{7}$$

$$U_p = \begin{cases} R_p i(t) \left[1 - e^{\frac{-(t-t_0)}{\tau_p}} \right] & t_0 < t \le t_d \\ U_p(t_d) e^{\frac{-(t-t_d)}{\tau_p}} & t_d < t < t_r \end{cases} \tag{8}$$

Let $\tau_s = R_s C_s$ and $\tau_p = R_p C_p$ be the time constants of the two parallel circuits. The voltage change $(V_3 - V_1)$ is caused by the disappearance of the polarization effect of the battery. In this process, the battery voltage output is $V = E - IR_s e^{\frac{-t}{\tau s}} - IR_p e^{\frac{-t}{\tau p}}$, which can be simply written as $V = E - ae^{-ct} - be^{-dt}$. This form can be used to fit the coefficients of the double exponential terms with MATLAB. After calculating a, b, c and d, according to $R_s = \frac{a}{I}$, $R_p = \frac{b}{I}$, $C_s = \frac{1}{R_s c}$, and $C_p = \frac{1}{R_p d}$, the values of R_s, R_p, C_s, and C_p can be identified.

2.2.3. Model Accuracy Verification

After the model parameters have been identified, the accuracy of the model needs to be verified to determine whether the model parameters can be used for SOC estimation and parameter optimization in this paper. The precision of the model is verified by referring to the hybrid pulse power characterization (HPPC) experiment [26]. The simulated initial SOC is set to 0.5. The comparison between the experimentally measured voltage and the model simulation output voltage is shown in Figure 7. To better distinguish the difference, the error is presented in Figure 8.

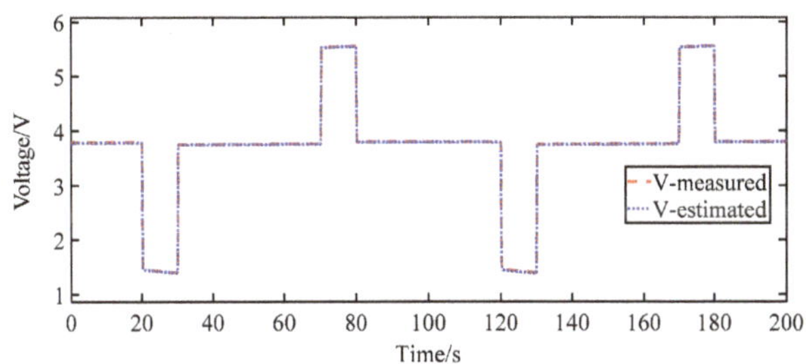

Figure 7. The voltage response comparison curve.

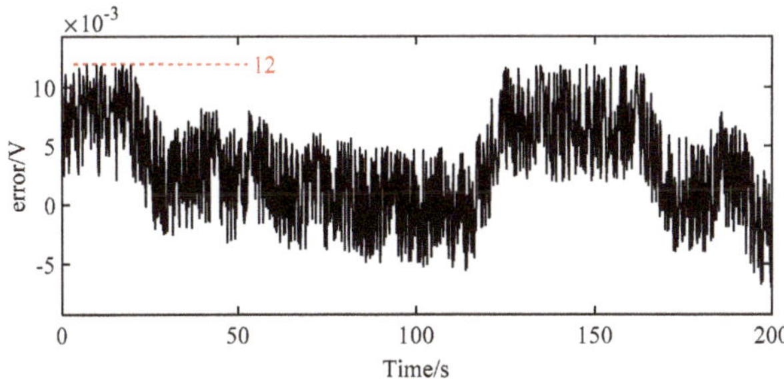

Figure 8. The voltage response error.

As seen from Figures 7 and 8, when the output voltage of the battery changes due to abrupt current changes, the output voltage of the simulation model can better track the measured voltage. The maximum error is 12 mV, which indicates that the model parameters can be used for the online SOC estimation of the Kalman filter and reasonable parameter optimization.

3. The AUKF Based on the Second-Order RC Equivalent Circuit Model

Based on the equivalent circuit model, the corresponding SOC estimation algorithm can be established. A power lithium battery is a typical nonlinear system. The nonlinear system is governed by the equations of state, and the observation equations are shown in Equation (9).

$$\begin{cases} x_k = f(x_{k-1}, u_{k-1}) + \omega_{k-1} \\ y_k = h(x_k, u_k) + v_k \end{cases} \tag{9}$$

where the random variables w_k and v_k represent the process and measurement noise, respectively. For the UKF, the iteration equation is based on a certain set of sample points, which are chosen to make their mean value and variance consistent with the mean value and variance of the state variables. Then, these points will recycle the equation of the discrete-time process model to produce a set of predicted points. After, the mean value and the variance of the predicted points will be calculated to modify the results, and the mean value and the variance will be estimated. Before the UKF recursion, the state variables must be modified in a superposition of the process noise and the measurement noise of the original states. The SOC of the lithium battery pack can be calculated using the ampere-hour integral method:

$$SOC(t) = SOC(t') - \frac{1}{C_N} \int_{t'}^{t} \eta i dt \tag{10}$$

In Equation (10), $\eta = \frac{k_i k_t}{k_c}$, where k_i is the compensation coefficient of the charge and discharge rate, k_t is the temperature compensation coefficient, k_c is the cycle compensation coefficient, and C_N is the actual available battery capacity. From the equivalent circuit model shown in Figure 3, the lithium battery equation of state can be obtained from Equations (2) and (10):

$$\begin{bmatrix} SOC_k \\ u_{s,k} \\ u_{p,k} \end{bmatrix} = \begin{bmatrix} 1 & 0 & 0 \\ 0 & a_s & 0 \\ 0 & 0 & a_p \end{bmatrix} \begin{bmatrix} SOC_{k-1} \\ u_{s,k-1} \\ u_{p,k-1} \end{bmatrix} + \begin{bmatrix} \frac{\eta T}{C_N} \\ b_s \\ b_p \end{bmatrix} I_{k-1} + \begin{bmatrix} \omega_1(k) \\ \omega_3(k) \\ \omega_5(k) \end{bmatrix} \tag{11}$$

$$\begin{aligned} U_k &= E_k - I_k R - U_{s,k} - U_{p,k} + v(k) \\ &= F(SOC_k) - I_k R - U_{s,k} - U_{p,k} + v(k) \end{aligned} \tag{12}$$

where the values of each coefficient are shown in Equation (5). For the circuit model shown in Equations (11) and (12):

$$X_k = \begin{bmatrix} x_k^T, & \omega_k^T, & v_k^T \end{bmatrix}^T = \begin{bmatrix} SOC_k, & u_{s,k}, & u_{p,k}, & \omega_{soc,k}, & \omega_{s,k}, & \omega_{p,k}, & v_k \end{bmatrix}^T \tag{13}$$

To facilitate the distinction, we can take $x_k = \begin{bmatrix} SOC_k, U_{s,k}, U_{p,k} \end{bmatrix}$ as the initial state of the system, y_k as the raw output (its corresponding symbol is U_k in the circuit model of the lithium battery), and u_k as the control variable (its corresponding symbol is I_k), and we can make $\Psi = [y_1, y_2, \cdots, y_k]$. The operations of an ordinary UKF are as follows [27–29]:

(1) Time update of state estimation

The mean and variance of the extended state are obtained based on the optimal state estimation at the previous time. Select $(2L + 1)$ sampling points (L is the dimension of the extended state, where $L = 7$). Finally, the sampling points are transformed by the state equation, and the state prediction is completed.

(1) Initialization, initial state determination:

$$\begin{cases} \hat{x}_0 = E[x_0] \\ P_0 = E\left[(x_0 - \hat{x}_0)(x_0 - \hat{x}_0)^T\right] \end{cases} \tag{14}$$

(2) State expansion:

$$\begin{cases} \hat{X}_0 = E[X_0] = E[\hat{x}_0,\ 0,\ 0] \\[2mm] P_{X,0} = E\left[\left(X_0 - \hat{X}_0\right)\left(X_0 - \hat{X}_0\right)^T\right] = \begin{bmatrix} P_0 & & \\ & Q & \\ & & R \end{bmatrix} \end{cases} \tag{15}$$

$$E[w_m,\ w_n] = \begin{cases} Q\ m = n \\ 0\ m \neq n \end{cases} \tag{16}$$

$$E[v_m,\ v_n] = \begin{cases} R\ m = n \\ 0\ m \neq n \end{cases} \tag{17}$$

Q and R are covariance matrices, which are symmetric, and the diagonal is the variance in each dimension.

State mean expansion:

$$\hat{X}_{k-1} = \left[\left(\hat{x}_{k-1}\right)^T, \overline{w}_k^T, \overline{v}_k^T\right]^T \tag{18}$$

State variance expansion:

$$P_{X,k-1} = \begin{bmatrix} P_{x,k-1} & & \\ & Q & \\ & & R \end{bmatrix} \tag{19}$$

(3) Sample point selection

Sample $= \{z_i, X_{k-1,i}\}$, $i = 0, 1, 2, \ldots, 2L + 1$, where $X_{k-1,i}$ represents the selected points, and z_i is the corresponding weighting value. Then, select the points in the following manner:

$$\begin{cases} X_{k-1,0} = \hat{X}_{k-1} \\[1mm] X_{k-1,i} = \hat{X}_{k-1} + \left(\sqrt{(L+\lambda)P_{X,k-1}}\right)_i & i = 1 \sim L \\[1mm] X_{k-1,i} = \hat{X}_{k-1} - \left(\sqrt{(L+\lambda)P_{X,k-1}}\right)_i & i = (L+1) \sim 2L \end{cases} \tag{20}$$

The corresponding weighting values are:

$$\begin{cases} z_0^{(m)} = \frac{\lambda}{L+\lambda} \\[1mm] z_0^{(c)} = \frac{\lambda}{L+\lambda} + \left(1 + \alpha^2 + \beta\right) \\[1mm] z_i^{(m)} = z_i^{(c)} = \frac{1}{2(L+\lambda)} & i = 1 \sim 2L \end{cases} \tag{21}$$

where $\lambda = \alpha^2(L+t) - L$, $z^{(m)}$ is the corresponding weighting value of the mean, $z^{(c)}$ is the corresponding weighting value of the variance, and $\left(\sqrt{(L+\lambda)P_{X,k-1}}\right)_i$ denotes the values of column i in the square-root matrix $(L+\lambda)P_{X,k-1}$. To ensure that the covariance matrix is definitely positive, we must take $t \geq 0$; α controls the distance of the selected points, with $10^{-2} \leq \alpha \leq 1$, and β is used to reduce the error of the higher-order terms. For a Gaussian [30], the optimal choice is $\beta = 2$ in this paper, along with $t = 0$ and $\alpha = 1$. $\hat{X}_{k-1,i}$ consists of $\hat{X}_{k-1,i}^x$, $\hat{X}_{k-1,i}^w$ and $\hat{X}_{k-1,i}^v$.

Based on this, the time update of the state estimation is performed:

$$\begin{aligned} \hat{x}_{k|k-1} &= E\{[f(x_{k-1},\ u_{k-1}) + w_{k-1}]|\Psi_{k-1}\} \\ &= \sum_{i=0}^{2L} z_i^{(m)}\left[A_{k-1}X_{k-1,i}^x + B_{k-1}u_{k-1} + X_{k-1,i}^w\right] \\ &= \sum_{i=0}^{2L} z_i^{(m)} X_{k|k-1,i}^x \end{aligned} \tag{22}$$

(2) Time update of mean square error:

$$
\begin{aligned}
P_{x,k|k-1} &= E\left[\left(x_k - \hat{x}_{k|k-1}\right)\left(x_k - \hat{x}_{k|k-1}\right)^T\right] \\
&= \sum_{i=0}^{2L} z_i^{(c)}\left(X_{k|k-1,i}^x - \hat{x}_{k|k-1}\right)\left(X_{k|k-1,i}^x - \hat{x}_{k|k-1}\right)^T
\end{aligned}
\tag{23}
$$

(3) A priori estimate of system output:

$$
\begin{aligned}
\hat{y}_k &= E\{[h(x_k,\,u_k) + v_k]|\Psi_{k-1}\} \\
&= \sum_{i=0}^{2L} z_i^{(m)}\left[h\left(X_{k|k-1}^x,\,u_k\right) + X_{k-1,i}^v\right] \\
&= \sum_{i=0}^{2L} z_i^{(m)} y_{k|k-1}
\end{aligned}
\tag{24}
$$

(4) Calculation of the filter gain matrix:

$$
\begin{aligned}
L_k &= P_{xy,k}P_{y,k}^{-1} \\
&= \sum_{i=0}^{2L} z_i^{(c)}\left(X_{k|k-1,i}^x - \hat{x}_{k|k-1}\right)\left(y_{k|k-1,i} - \hat{y}_k\right)^T\left[\sum_{i=0}^{2L} z_i^{(c)}\left(y_{k|k-1,i} - \hat{y}_k\right)\left(y_{k|k-1,i} - \hat{y}_k\right)^T\right]^{-1}
\end{aligned}
\tag{25}
$$

(5) Optimal state estimation:

$$
x_k = \hat{x}_{k|k-1} + L_k(y_k - \hat{y}_k)
\tag{26}
$$

where y_k is the actual measured value of the system output.

(6) Estimation of mean square error:

$$
P_{x,k} = P_{x,k|k-1} - L_k P_{y,k} L_k^T
\tag{27}
$$

As the process noise and measurement noise are measured in real-time, and to update the covariance of the process noise and measurement noise in real time, the following is needed:

$$
\begin{cases}
\mu_k = y_k - h[\hat{x}_k, u_k] \\
F_k = \mu_k \mu_k^T \\
R_k^v = \left(F_k + \sum_{i=0}^{2L} z_i^c\left(y_{k|k-1,i} - y_k\right)\left(y_{k|k-1,i} - y_k\right)^T\right)/2 \\
R_k^w = L_k F_k L_k^T
\end{cases}
\tag{28}
$$

where μ_k is the residual error of the system measured output, and $y_{k|k-1,i}$ is the residual error of the system measured output estimated by the sigma points. Real-time updates of the process noise and measurement noise can be achieved by Equation (28). Thus, the establishment of the AUKF is completed.

4. Reasonable Optimization of Model Parameters

In the process of identifying the parameters of the battery model, it was found that the amount of experimental data is always limited for a battery that works continuously for a period of time. When these raw data are substituted into the model for SOC estimation, the parameter values before and after each measured model parameter are prone to abrupt changes.

This paper reasonably optimizes the model parameters when the SOC is 0.1 X (where X is an integer from 1 to 9). In Figure 9, the ohmic resistance R_0 at the time when the experimentally measured SOC is equal to an integer multiple of 0.1 is shown.

When these raw data are substituted into the model for SOC estimation, there will be more abrupt changes or "spikes" in the variation curve of the parameters when the SOC is equal to an integer multiple of 0.1, as shown by the black circle in Figure 9. However, in actual continuous operation conditions, the change trend of each parameter of the battery is continuous and relatively flat, without an obvious inflection point. To make the data curve as flat as possible on the premise of approaching the real value, reasonable methods are considered to optimize the data measured in the test to eliminate the spike. Conventional polynomial fitting results in a very smooth fitting curve, but often has a large deviation from the true value, and the error trend at both ends of the curve is the most serious.

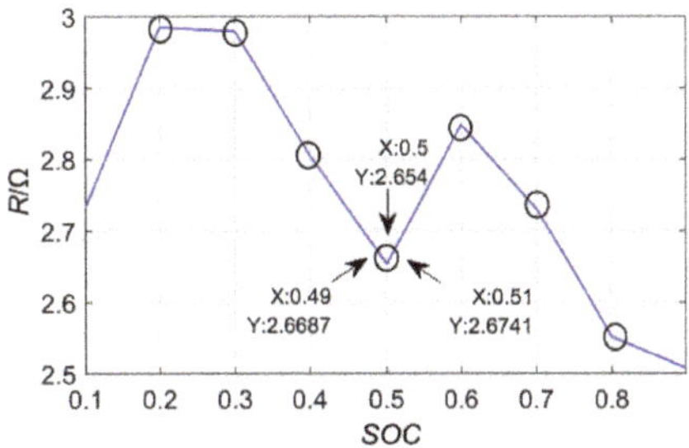

Figure 9. Original data curve.

Given the data accuracy and the slowly changing curve characteristics, this paper adopts a reasonable optimization method to weaken data mutation. In time, $SOC = 0.1X(X = 1, 2 \ldots \ldots 9)$, the model parameters are the identified real values, and the model parameters are optimized before and after 0.1 X. That is, point $(0.1X - 0.01)$ and point $(0.1X + 0.01)$ are selected around point $SOC = 0.1X$ to optimize the model parameters of the point $SOC = 0.1X$. When $SOC = (0.1X - 0.01)$, the model parameter values at times $SOC = 0.1X$ and $SOC = (0.1X - 0.1)$ are connected by a straight line, and the model parameter value at time $SOC = (0.1X - 0.01)$ is obtained by the functional relationship of the straight line and set as m. When $SOC = (0.1X + 0.01)$, the model parameter values when $SOC = 0.1X$ and $SOC = (0.1X + 0.1)$ are connected by a straight line, and the model parameter value at time $SOC = (0.1X + 0.01)$ is obtained as n according to the functional relationship of the straight line. $(m + n)/2$ is taken as the optimal value of the model parameter at time $SOC = 0.1X$. After optimization, there is a small error in the model parameters at $SOC = 0.1X$, but, in return, the model parameters are closer to the curve of continuous changes in the actual working conditions.

The data curve of the battery internal resistance $R0$ measured at 25 °C and $SOC = 0.5$ is taken as an example. The $R0$ values of group $SOC = 0.49$ and group $SOC = 0.51$ are selected as model reference data before and after the mutation point, and the average value of the times $SOC = 0.49$ and $SOC = 0.51$ are calculated as the analysis values of the original "peak" point. The obtained optimized data curve is shown in Figure 10.

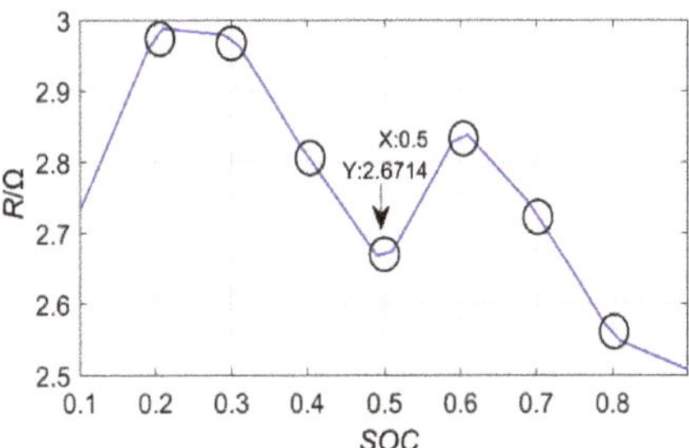

Figure 10. Optimized data curve.

From the optimized data curve, it can be seen that each data mutation and spike is weakened or smoothed, making the change trend of each parameter of the model closer to the actual operating conditions of the EVs. Similarly, the rest of the model parameter data are reasonably optimized (where the V_{oc} original data curve is relatively smooth and will not be optimized).

5. Comparison of Online SOC Estimation before and after Model Parameter Optimization

The verification of the AUKF based on the optimization of the model parameters is divided into three aspects. First, the estimation accuracy of the AUKF is compared to that of the normal UKF. Second, the superiority of the AUKF estimation accuracy before and after the optimization of the model parameters is verified. Finally, the convergence of the AUKF to the initial value error of the SOC before and after the optimization of the model parameters is verified.

In the process of setting the battery charge and discharge state, the UDDS (Urban Dynamometer Driving Schedule) waveform [27] is used as a reference. To ensure correspondence with the experimental object (SANYO 18650 Li-ion battery) in the process of OCV–SOC calibration, the current signal in Figure 11 is adopted to describe the increase or decrease of the current in the discharging or charging process of the power battery. In one period, the average output current is 1.77 A, the maximum discharging current is 5.28 A, and the maximum charging current is 2.42 A. Each period is 1367 s, and the condition lasts for two periods.

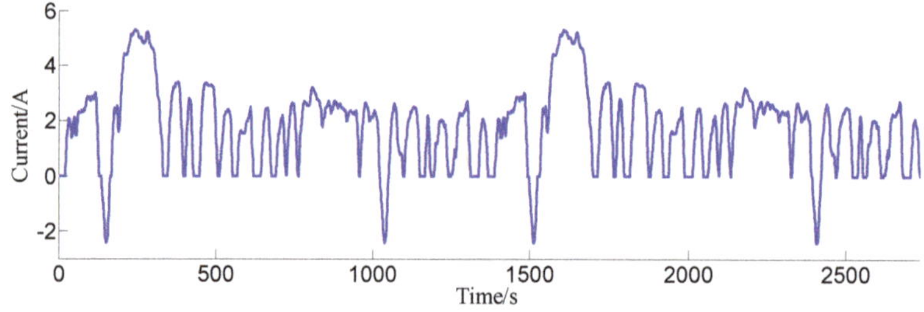

Figure 11. The input current waveform.

5.1. Comparison of the Estimation Accuracy between the AUKF and the UKF

The third section explains in detail the process of establishing the AUKF. This section verifies the superiority of the AUKF and the UKF in terms of estimation accuracy in MATLAB/Simulink. The verification process is based on the model parameters before optimization. In this simulation model, the input current $I(k)$ is integrated using the ampere-hour integral method. As there is no error in the current measurement due to outside disturbances, no accumulative error exists. Thus, the integration of the current in the simulation model can be regarded as the theoretical value of the SOC. The model simulation results and errors are shown in Figures 12 and 13, respectively.

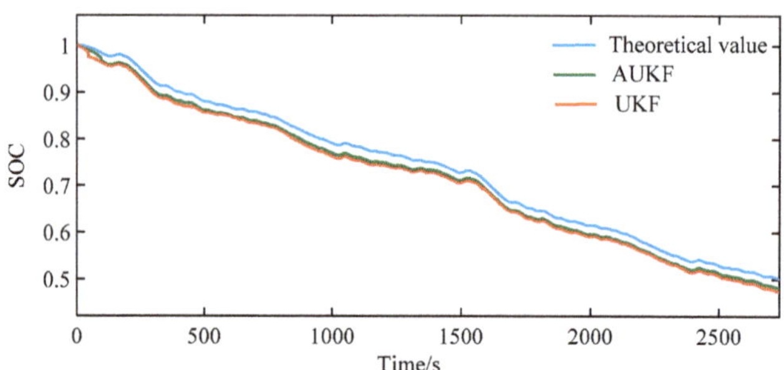

Figure 12. The model simulation results.

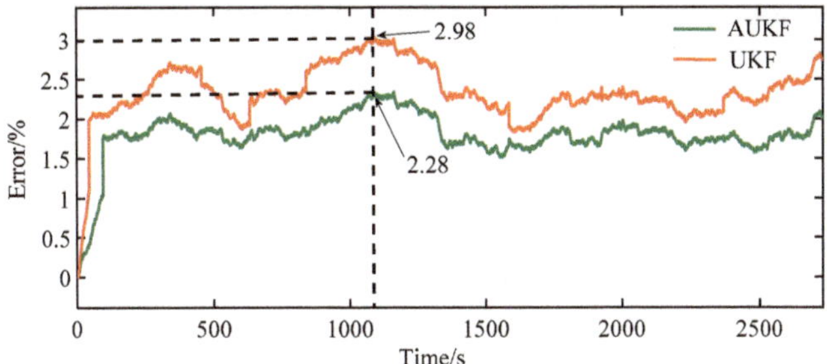

Figure 13. The model simulation errors.

From the simulation results, the AUKF and the UKF can estimate the SOC online well under different charge and discharge rates. The UKF maximum estimation error is 2.98%, and the maximum estimated error of the AUKF is 2.28%; compared with the UKF, its estimation accuracy is significantly improved.

5.2. Comparison of the AUKF Estimation Accuracy before and after Optimization of the Model Parameters

Substituting the optimized model parameters into the AUKF simulation model, the results are compared with the estimated values before the optimization of the model parameters. The simulation output is shown in Figure 14, and the error is shown in Figure 15.

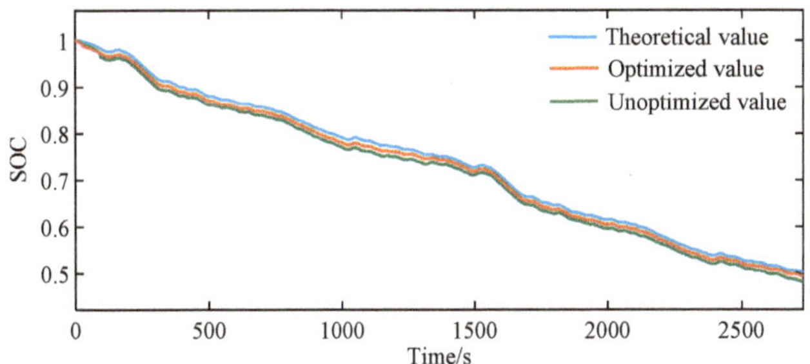

Figure 14. SOC estimation before and after optimization of the model parameters.

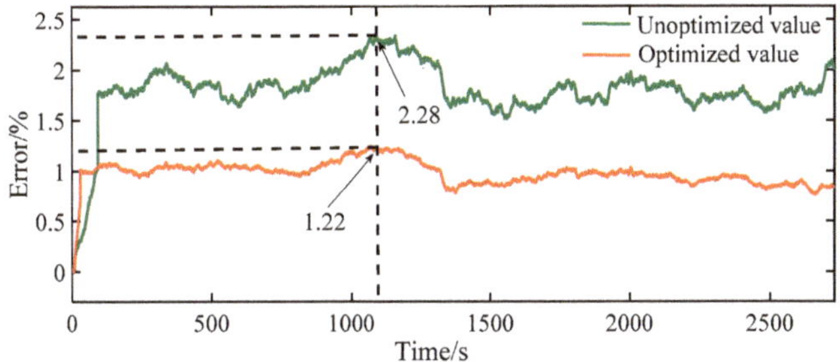

Figure 15. SOC estimation error before and after optimization of the model parameters.

From the simulation results shown in Figures 14 and 15, the online estimation of the SOC is closer to the theoretical value after the reasonable optimization of the model parameters. The maximum SOC estimation error obtained after optimization of the model parameters is 1.22% and compared to the maximum error of 2.28% before parameter optimization, the error decreases significantly. It is verified that the optimization process can significantly improve the online estimation accuracy of the AUKF based on the original data.

5.3. Convergence Comparison of the Initial Error of the AUKF before and after Parameter Optimization

The Kalman filter has a strong convergence effect on the initial error of the SOC. This section verifies the effect of the reasonable optimization of the model parameters on the initial error convergence ability of the AUKF. The initial value of the system state SOC in the AUKF is set to the same value, and the output of the AUKF estimation before and after model parameter optimization is compared. The initial SOC value is 0.9, and the simulation lasts for one cycle. A waveform comparison between the SOC value of the simulation model and the theoretical value is shown in Figure 16. Figure 17 shows the convergence of the initial error before and after the model parameter optimization of the first 400 s. The SOC estimation values before and after optimization at 200 s are given.

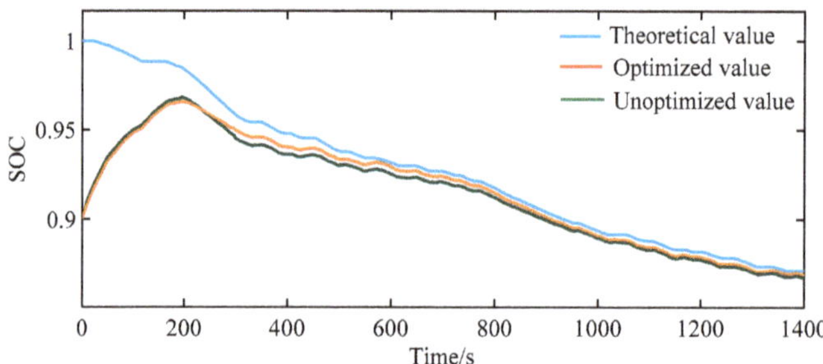

Figure 16. The comparison of the SOC estimation between the simulated value and the theoretical value.

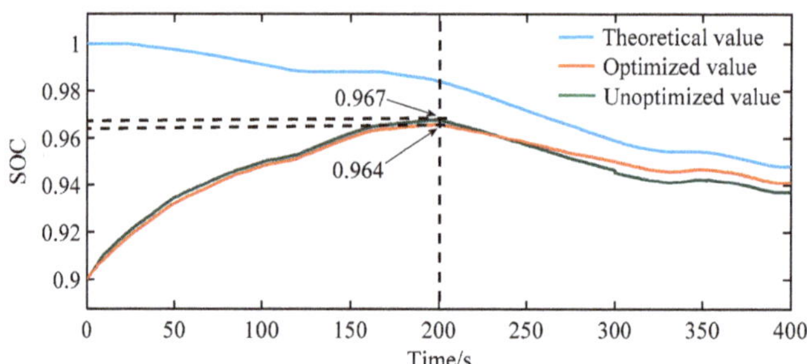

Figure 17. The comparison of the estimation values in the first 400 s.

As seen from Figure 16, before and after model parameter optimization, the SOC values estimated by the AUKF converge to the theoretical value quickly. In Figure 17, at 200 s, the estimated value before optimization is 0.967, which is closer to the theoretical value than the optimized estimated value of 0.964, but the difference is only 0.003; that is, the difference is only 0.3%, which can be ignored. This shows that the convergence ability of the AUKF algorithm to the initial error before and after optimization of the model parameters is approximately the same, and optimization of the model parameters has no effect on the convergence of the algorithm.

6. Conclusions

The SOC provides vital state information for EVs and is strongly nonlinear and time-varying. For the purpose of the accurate online estimation of the SOC, the parameters of a second-order RC equivalent circuit model are identified by combining circuit principles and optimized rationally.

The AUKF with optimized model parameters is used to estimate the SOC. The conclusions of this work can be summarized as follows:

1. The second-order RC model is proven to be the most suitable equivalent circuit model compared with the Thevenin and multi-order models.
2. The covariances of the process noise and the measurement noise of the UKF are updated in real time to improve the accuracy of SOC estimation.
3. A reasonable optimization method is adopted to weaken model parameter mutation. Optimized parameters can significantly improve the online estimation accuracy of the AUKF based on the original data and have little effect on the convergence of the algorithm.

In the future, to further improve the accuracy of SOC estimation, the influence of temperature and SOH on SOC estimation must be considered.

Author Contributions: Conceptualization, X.G. and X.X.; methodology, X.G.; software, X.G.; validation, J.G., Y.G. and X.X.; formal analysis, Z.L.; investigation, X.G.; resources, X.H.; data curation, X.H.; writing—original draft preparation, X.G.; writing—review and editing, X.X. and J.G.; visualization, X.X.; supervision, X.G.; project administration, X.G. and X.X.; funding acquisition, X.G. and X.X.

Funding: This research was funded by the National Natural Science Foundation of China, grant number 61703145. The Key Scientific Research Projects of Higher Education Institutions in Henan Province, grant number 19A470001, 18A470014. The Key Scientific and Technological Research Projects in Henan Province, grant number 192102210073. The APC was funded by 192102210073.

Acknowledgments: I would like to express my deepest gratitude to my supervisor, Longyun Kang, who has provided me with valuable guidance at every stage of writing this paper. I would also like to thank the anonymous reviewers for dedicating the time to review my paper despite their busy schedules.

Conflicts of Interest: The authors declare no conflict of interest.

References

1. Tong, S.; Klein, M.P.; Park, J.W. On-line optimization of battery open circuit voltage for improved state-of-charge and state-of-health estimation. *J. Power Sources* **2015**, *293*, 416–428. [CrossRef]
2. Pan, H.; Lü, Z.; Lin, W.; Li, J.; Chen, L. State of charge estimation of lithium-ion batteries using a grey extended Kalman filter and a novel open-circuit voltage model. *Energy* **2017**, *138*, 764–775. [CrossRef]
3. Guo, X.; Kang, L.; Yao, Y.; Huang, Z.; Li, W. Joint estimation of the electric vehicle power battery state of charge based on the least squares method and the Kalman filter algorithm. *Energies* **2016**, *9*, 100. [CrossRef]
4. Yang, F.; Xing, Y.; Wang, D.; Tsui, K.-L. A comparative study of three model-based algorithms for estimating state-of-charge of lithium-ion batteries under a new combined dynamic loading profile. *Appl. Energy* **2016**, *164*, 387–399. [CrossRef]
5. Meng, J.; Luo, G.; Ricco, M.; Swierczynski, M.; Stroe, D.I.; Teodorescu, R. Overview of lithium-ion battery modeling methods for state-of-charge estimation in electrical vehicles. *Appl. Sci.* **2018**, *8*, 659. [CrossRef]
6. Xiong, R.; Cao, J.; Yu, Q.; He, H.; Sun, F. Critical review on the battery state of charge estimation methods for electric vehicles. *IEEE Access* **2017**, *6*, 1832–1843. [CrossRef]
7. Meng, J.; Ricco, M.; Luo, G.; Swierczynski, M.; Stroe, D.I.; Stroe, A.I.; Teodorescu, R. An overview and comparison of online implementable SOC estimation methods for lithium-ion battery. *IEEE Trans. Ind. Appl.* **2017**, *54*, 1583–1591. [CrossRef]
8. Zheng, Y.; Ouyang, M.; Han, X.; Lu, L.; Li, J. Investigating the error sources of the online state of charge estimation methods for lithium-ion batteries in electric vehicles. *J. Power Sources* **2018**, *377*, 161–188. [CrossRef]
9. Hannan, A.; Lipu, M.S.H.; Hussain, A.; Mohamed, A. A review of lithium-ion battery state of charge estimation and management system in electric vehicle applications: Challenges and recommendations. *Renew. Sustain. Energy Rev.* **2017**, *78*, 834–854. [CrossRef]
10. Lai, X.; Qiao, D.; Zheng, Y.; Zhou, L. A fuzzy state-of-charge estimation algorithm combining ampere-hour and an extended kalman filter for Li-ion batteries based on multi-model global identification. *Appl. Sci.* **2018**, *28*, 2028. [CrossRef]

11. Ismail, M.; Dlyma, R.; Elrakaybi, A.; Ahmed, R.; Habibi, S. Battery state of charge estimation using an Artificial Neural Network. In Proceedings of the 2017 IEEE Transportation Electrification Conference and Expo (ITEC), Chicago, IL, USA, 22–24 June 2017.
12. Bishop, G.; Welch, G. An Introduction to the Kalman Filter. *Proc SIGGRAPH* **2001**, *8*, 41.
13. Julier, S.J.; Uhlmann, J.K. New extension of the Kalman filter to nonlinear systems. *Signal Process. Sens. Fusion Target Recognit. VI* **1999**, *3068*, 182–193.
14. Xiong, R.; Zhang, Y.; He, H.; Zhou, X.; Pecht, M.G. A double-scale, particle-filtering, energy state prediction algorithm for lithium-ion batteries. *IEEE Trans. Ind. Electron* **2017**, *65*, 1526–1538. [CrossRef]
15. Sepasi, S.; Ghorbani, R.; Liaw, B.Y. Improved extended Kalman filter for state of charge estimation of battery pack. *J. Power Sources* **2014**, *255*, 368–376. [CrossRef]
16. Pérez, G.; Garmendia, M.; Reynaud, J.F.; Crego, J.; Viscarret, U. Enhanced closed loop State of Charge estimator for lithium-ion batteries based on Extended Kalman Filter. *Appl. Energy* **2015**, *155*, 834–845. [CrossRef]
17. Li, Z.; Zhang, P.; Wang, Z.; Song, Q.; Rong, Y. State of Charge Estimation for Li-ion Battery Based on Extended Kalman Filter. *Energy Procedia* **2017**, *105*, 3515–3520.
18. Han, W.; Zou, C.; Zhou, C.; Zhang, L. Estimation of Cell SOC Evolution and System Performance in Module-based Battery Charge Equalization Systems. *IEEE Trans. Smart Grid* **2018**, *10*, 4717–4728. [CrossRef]
19. Wei, H.; Zhang, L.; Han, Y. Computationally efficient methods for state of charge approximation and performance measure calculation in series-connected battery equalization systems. *J. Power Sources* **2015**, *286*, 145–158.
20. Partovibakhsh, M.; Liu, G. An Adaptive Unscented Kalman Filtering Approach for Online Estimation of Model Parameters and State-of-Charge of Lithium-Ion Batteries for Autonomous Mobile Robots. *IEEE Trans. Control Syst. Technol.* **2014**, *23*, 357–363. [CrossRef]
21. Cai, M.; Chen, W.; Tan, X. Battery state-of-charge estimation based on a dual unscented Kalman filter and fractional variable-order model. *Energies* **2017**, *10*, 1577. [CrossRef]
22. Guo, X. Study on State of Charge Estimation and Equalization Technique of Electric Vehicle Battery. Ph.D. Thesis, South China University of Technology (SCUT), Guangzhou, China, 2016.
23. Xikun, C.; Dong, S. Research on lithium-ion battery modeling and model parameter identification methods. *Proc. CSEE* **2016**, *36*, 6254–6261.
24. Zhichao, H.; Geng, Y.; Languang, L. A Modeling for Power Battery Dynamics. *Trans. Electrotech. Soc.* **2016**, *31*, 194–203.
25. He, H.; Zhang, X.; Xiong, R.; Xu, Y.; Guo, H. Online model-based estimation of state-of-charge and open-circuit voltage of lithium-ion batteries in electric vehicles. *Energy* **2012**, *39*, 310–318. [CrossRef]
26. Ranjbar, A.H.; Banaei, A.; Khoobroo, A.; Fahimi, B. Online Estimation of State of Charge in Li-Ion Batteries Using Impulse Response Concept. *IEEE Trans. Smart Grid* **2012**, *3*, 360–367. [CrossRef]
27. Yu, Q.; Xiong, R.; Lin, C.; Shen, W.; Deng, J. Lithium-ion battery parameters and state-of-charge joint estimation based on H-infinity and unscented Kalman filters. *IEEE Trans. Veh. Technol.* **2017**, *66*, 8693–8701. [CrossRef]
28. Li, K.; Chang, L.; Hu, B. A variational Bayesian-based unscented Kalman filter with both adaptivity and robustness. *IEEE Sens. J.* **2016**, *16*, 6966–6976. [CrossRef]
29. Li, Y.; Wang, C.; Gong, J. A wavelet transform-adaptive unscented kalman filter approach for state of charge estimation of LiFePo4 battery. *Int. J. Energy Res.* **2018**, *42*, 587–600. [CrossRef]
30. Kandepu, R.; Foss, B.; Imsland, L. Applying the unscented Kalman filter for nonlinear state estimation. *J. Process Control* **2008**, *18*, 753–768. [CrossRef]

Article

Cooperative Optimization of Electric Vehicles and Renewable Energy Resources in a Regional Multi-Microgrid System

Jin Chen, Changsong Chen * and Shanxu Duan

State Key Laboratory of Advanced Electromagnetic Engineering and Technology, School of Electrical and Electronic Engineering, Huazhong University of Science and Technology, Wuhan 430074, China; eufoniuso@163.com (J.C.); duanshanxu@hust.edu.cn (S.D.)
* Correspondence: ccsfm@163.com

Received: 18 April 2019; Accepted: 27 May 2019; Published: 31 May 2019

Featured Application: The main goal of this work is to provide a cooperative optimization method for regional multi-microgrid system, optimizing dispatching strategy and capacity allocation of electric vehicles and renewable energy sources. Meanwhile, across-time-and-space energy transmission of electric vehicles is considered in the optimization model.

Abstract: By integrating renewable energy sources (RESs) with electric vehicles (EVs) in microgrids, we are able to reduce carbon emissions as well as alleviate the dependence on fossil fuels. In order to improve the economy of an integrated system and fully exploit the potentiality of EVs' mobile energy storage while achieving a reasonable configuration of RESs, a cooperative optimization method is proposed to cooperatively optimize the economic dispatching and capacity allocation of both RESs and EVs in the context of a regional multi-microgrid system. An across-time-and-space energy transmission (ATSET) of the EVs was considered, and the impact of ATSET of EVs on economic dispatching and capacity allocation of multi-microgrid system was analyzed. In order to overcome the difficulty of finding the global optimum of the non-smooth total cost function, an improved particle swarm optimization (IPSO) algorithm was used to solve the cooperative optimization problem. Case studies were performed, and the simulation results show that the proposed cooperative optimization method can significantly decrease the total cost of a multi-microgrid system.

Keywords: electric vehicles; renewable energy sources; microgrid; economic dispatching; capacity allocation; cooperative optimization

---

1. Introduction

Due to the characteristics of low carbon emission and sustainability, renewable energy sources (RESs) have attracted much attention in recent years [1–3]. In order to cope with the intermittency and uncertainty of RESs, the concept of microgrid (MG) was proposed to increase the utility of RESs by integrating them with energy storage units [4]. However, available storage units are expensive and a massive usage of them would significantly add to the cost of operation. The complexity of a microgrid's controlling is also an issue which challenges the traditional ways a power system operates.

However, increasing of electric vehicles' (EV) penetration brings more uncertainty to the operation of a power system. Uncontrolled charging behaviors of EV owners in load peak hours aggravate the burden of a power grid, posing great threats to the grid's operation.

Since EVs have the ability of storing energy, they can serve as storage units in RES-equipped power systems. Taking EVs as energy storage units and integrating them with RESs can attenuate the burden of disorderly charging while reducing the installation of expensive storage batteries [5].

Many studies focusing on the integration of EVs and RESs have been published in recent years; the main topics of these can be briefly separated into two categories, economic dispatching and system capacity allocation.

Most research has discussed the issue from the aspect of economic dispatching. The majority of the studies elaborated the optimal charging management of EVs [6,7]. However, as a kind of storage units, EVs can also discharge to the main grid or microgrid, providing a vehicle-to-grid (V2G) service to participate into the operation of the power systems [8,9]. Dispatching EVs which provide V2G services was also discussed in many papers. In [10], a MILP (Mixed-integer linear programming) model is proposed to find the optimal charging and discharging strategy of hybrid electric vehicles (HEVs). The proposed method in the paper improved the automation of HEV navigation systems and find a way to a more economic and sustainable transportation system. In [11], an integrated microgrid system considering both renewable energy source and electric vehicles is constructed. EV owners are regarded as a kind of flexible load of the demand response (DR), and the proposed optimization method significantly reduced the cost of the system. Though there is a large number of works that focus on the V2G service of EVs, some of them achieving significant progress, there are still some issues to be addressed. Most models proposed in the works are applied on only single microgrids. Interaction among different microgrids is not considered. On the other hand, the mobility of EVs is not fully exploited, and the across-time-and-space energy transmission of EVs in not considered.

Capacity allocation of the integration system is also an important issue in the optimization of the system, but not frequently mentioned in research. Existing works focus only on the planning of BESSs (Battery Energy Storage Systems) and RESs. Most of the research allocated either charging devices or RESs rather than cooperatively optimizing them at the same time. In [12], planning different types of plug-in electric vehicles (PEV) that charge infrastructures were studied. On the other hand, in [13], capacities of RESs were optimized to minimize the whole system's investment cost while meeting the load demand. In [14], an algorithm for microgrid planning was proposed considering massive connection of EV charging demands, and the system investment cost as well as CO_2 emission is reduced. However, on the one hand, the relationship between economic dispatching and capacity configuration of the microgrid system is not considered in the papers. On the other hand, only a few of the studies discuss the allocation for both RESs and EVs, and the concept of cooperative optimization is not found in any research.

To sum up, there has already been much research on the topic of optimal dispatching and system allocation of RES-EV integrated systems [15,16]. Nevertheless, EV's ability of across-time-and-space energy transmission (ATSET) in the context of multi-microgrid systems is not discussed in the literature. The relationship between system economic dispatching and capacity configuration is not considered, and cooperative optimization method considering both allocation and dispatching are not mentioned in the studies. On the one hand, an allocation without system dispatching would cause a redundant installation of RESs and increase the total cost of the system. On the other hand, EV dispatching without optimal allocation of the system would decrease the efficiency of their integration into the microgrid system. Therefore, a cooperative optimization that simultaneously considers allocation and dispatching is necessary for the system's economic operation. For the reference, an optimization method using both the configuration of RESs and EVs is not considered in most of the works.

In order to fully exploit the potential of EV's mobile storage ability as well as reduce the redundant installation of RESs, a cooperative optimization considering EV's across-time-and-space energy transmission is presented in this paper. Both the installation capacity of RESs and the number of EV charging/discharging infrastructures (EVCDIs) are considered as allocation optimization variables.

The main contributions of this paper are listed as follow:

1. An integration mechanism of RESs and EVs is illustrated, and a mathematical model of the V2G-integrated regional multi-microgrid system is established;
2. The concept of EV's ATSET is illustrated. Based on the ability of EV's ATSET, a cooperative optimization method for economic dispatching and system capacity allocation is proposed;

3. To fully exploiting the potential of the EV's mobile energy storage ability, both the installation capacity of RESs and the number of EVCDIs are optimized in the cooperative optimization process.

The following manuscript is organized as follows. In Section 2, basic concept of cooperative optimization is illustrated. Mathematical model of a typical regional multi-microgrid system is constructed and elaborated in Section 3. In Section 4, objective functions and restrictions are described. The cooperative optimization model is demonstrated in Section 5 and case studies are performed in Section 6. Finally, in Section 7, conclusions are drawn.

2. Concept of Cooperative Optimization

The theoretical structure of the cooperative optimization method proposed in this paper can be illustrated as in Figure 1.

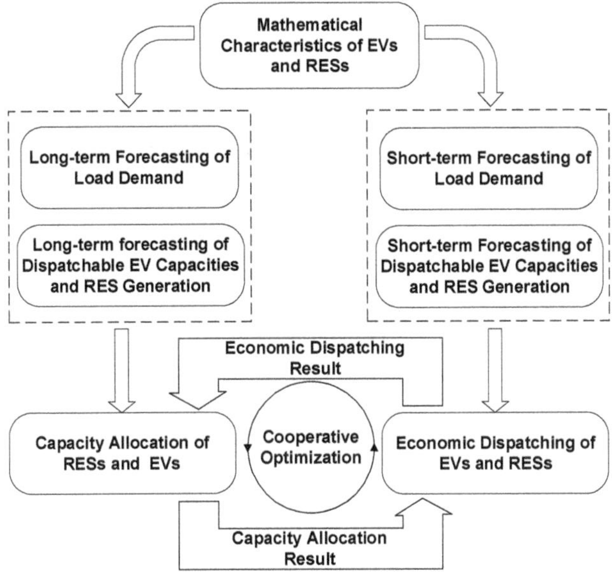

Figure 1. Brief process of cooperative optimization of EVs (Electric Vehicles) and RESs (Renewable Energy Sources).

The model can be separated into two sections: one is the forecasting of load demands and RES generations—the accuracy of the forecast can significantly influence the availability of the optimization results—and the other is the main part of the cooperative optimization. In the optimization process, forecasting data are derived according to the mathematical character of EVs and RESs, sent to the dispatching and allocation sections. The dispatching model will dispatch the operation of microgrid system including EVs and RESs, and the allocation section can optimize the installation capacities of EVBCDIs (Electric Vehicle bidirectional charging/discharging infrastructure) and RESs. An allocation module generates the initial system allocation. A dispatching model then optimizes the operation of the system according to the system allocation and feeds the results back to the allocation module. The system allocation is then updated by an allocation module according to the data from the dispatching module. Through several repetitions of this interaction, both allocations and dispatching of the system are finally optimized. Details of the cooperative optimization process will be discussed in following sections.

In this paper, three respects of cooperative optimization are considered.

Firstly, RESs and EVs are cooperatively optimized. By optimally dispatching EVs, energy generated by RESs can be redistributed across different times and spaces, and the utilization efficiency of RESs can be improved. Furthermore, EVs can also profit from interacting with RESs and participating into the operation of the system.

Secondly, microgrids in the regional multi-microgrid system are cooperatively optimized. With the ATSET of EVs, energy can be transmitted through different microgrids, and microgrids with lower electricity prices indirectly sell their energy to microgrids with higher electricity prices. Through such a cooperation, microgrids with lower prices can sell energy to make profits, and microgrids with higher prices can purchase electricity form EVs with lower prices.

Thirdly, economic dispatching and system allocation are cooperatively optimized. An economic dispatching of the system can reduce the cost of a redundant installation of RESs and EVCDIs, and an appropriate system allocation can in turn better exploit the potential of generation and energy storage units to serve the load with less costs.

3. Mathematical Model

3.1. Structure of the Multi-Microgrid System

The structure of a classic multi-microgrid system is demonstrated as follows in Figure 2. The system is constructed with several independent microgrids, which can be briefly categorized into residential microgrids (RMGs) and office building microgrids (OBMGs). The load curves of these two kinds of microgrids are slightly different, and OBMGs have a heavier load demand in general.

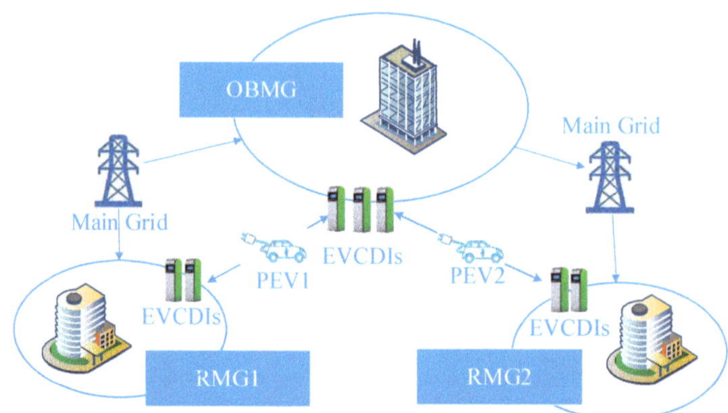

Figure 2. Structure of a classic multi-microgrid system.

In every single microgrid, photovoltaic (PV) generation modules and domestic small-size wind turbines (WTs) are installed according to the microgrid's load demand. To satisfy the requests for charging of both residents and staff, charging plots were installed in all microgrids, and parts of them include EVBCDIs for participating in the power system's operation. For the sake of power balance and the stability of system's operation, all microgrids are connected to the main grid through connection transformers.

3.2. Renewable Energy Sources

Most load demand of microgrid systems are supplied by distributed energy sources, the majority of which are the RESs mentioned in this paper. Since RESs' characters are highly relevant to the weather and environment, their generations are volatile and intermittent. For the sake of safety and availability, an accurate and reliable prediction of RES generation is necessary. In this paper, artificial neural networks (ANNs) are adopted as the prediction model, and two models from [17,18] are used in the proposed optimization model. Due to the restriction of the paper length, details of the two models are not demonstrated here.

3.3. Electric Vehicles

The multi-microgrid system considered in this paper is a living–working combined system. Residents are all staff from office building areas. Thus, all the EVs considered in this paper are commuting cars. They park in the RMG to charge during the knock-off hours from 19:00 to 8:00 and park in the OBMG during working time from 9:00 to 18:00. Considering the random behaviors of the staff, the arrival and departing times of EVs were derived by Monte Carlo experiment and random variables were then added.

Energy variation of EV batteries came from two aspects. One is the energy consumption of EVs for their driving on the commuting route, and the other is charging and discharging through EVBCDIs. State of capacity (SOC) was used to present the energy remaining in the mobile batteries. During the driving distance, the SOC of PEVs could be calculated as (1).

$$SOC_{i,j}^{EV}(t) = SOC_{i,j}^{EV}(t - d_{i,j}) - D_{i,j}^{EV} \cdot C_d \tag{1}$$

where $SOC_{i,j}^{EV}(t)$ is the SOC of the jth EV in the ith microgrid in the tth hour, and $d_{i,j}$ is the driving time. C_d is the energy consumption rate of EV's driving, and $D_{i,j}^{EV}$ is the driving distance. Since EV's driving distances are slightly different due to each owner's random behaviors, the distance was also generated from Monte Carlo experiment, and can be demonstrated as Equation (2) [19].

$$D_{i,j}^{EV} = E(D_{i,j}^{EV}) + k_{i,j} \cdot \sigma_{i,j}^{EV} \tag{2}$$

where $D_{i,j}^{EV}$ is the actual value of $D_{i,j}^{EV}$; $E(D_{i,j}^{EV})$ is the mathematical expectation of the driving distance; $k_{i,j}$ is a random parameter and is normally distributed; $\sigma_{i,j}^{EV}$ is the standard deviation of the driving distance. Notice that according to the statistical data of automobile's driving, the driving distance is also normally distributed.

When parking in the OBMG or RMG, the SOCs of EVs were determined only by the charging and discharging behaviors. Considering the self-discharging of EV batteries, the SOC can be calculated according to Equation (3).

$$\begin{cases} SOC_{i,j}^{EV}(t) = SOC_{i,j}^{EV}(t-1) \cdot (1 - \delta_{SD}) \\ \quad\quad -P_{i,j}^{EV}(t) \cdot \Delta t \cdot \eta_C & if\ P_{i,j}^{EV}(t) < 0 \\ SOC_{i,j}^{EV}(t) = SOC_{i,j}^{EV}(t-1) \cdot (1 - \delta_{SD}) \\ \quad\quad -P_{i,j}^{EV}(t) \cdot \Delta t / \eta_D & if\ P_{i,j}^{EV}(t) > 0 \end{cases} \tag{3}$$

where $P_{i,j}^{EV}(t)$ is the charging and discharging power of the jth EV in the ith microgrid in the tth hour, while $P_{i,j}^{EV}(t) < 0$ for charging and $P_{i,j}^{EV}(t) > 0$ for discharging. δ_{SD} is the self-discharging rate. η_C and η_D are the charging and discharging efficiency rates, respectively.

3.4. Operation Mechanism

As illustrated in Section 3.1, the system is mainly formulated by EVs, RESs, EVBCDIs, and other assistant devices as connection transformers. In order to optimize the system's operation, data of SOC are collected by the intelligent meters installed in the EVBCDIs and then uploaded to the microgrid management system (MMS) and finally to the general management system (GMS) of the whole multi-microgrid system. The GMS optimizes system operation data and then send the dispatching results to the EVBCDIs to be performed. A brief description of the operation mechanism for the cooperative optimization is demonstrated as Figure 3.

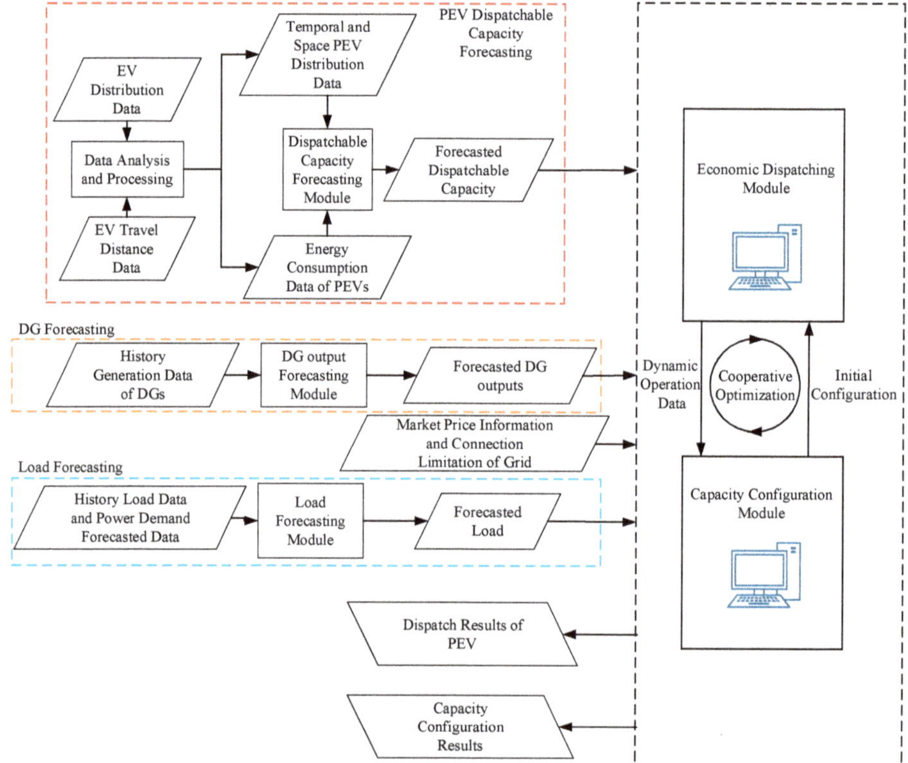

Figure 3. Mechanism of the cooperative optimization.

Since there are many EVs in each microgrid, it is not impractical to optimize the charging and discharging powers of every EV, nor is it necessary. In this paper, a hierarchical structure of the management system is utilized. EVs are separated into several EV fleets according to the residential areas they belong to, and each fleet is consisted of all the EVs form the same RMG. Charging and discharging powers of EV fleets rather than single EVs are optimized by GMS, and the data are sent to the management systems of EV fleets. The power of single EVs is then optimized according to the data from the GMS. In this paper, only the optimal dispatching of EV fleets is discussed, and the dispatching method of single EVs can refer to [20] and is not comprehensively discussed in this work.

4. Optimization Problem Formulation

As a typical mathematical programming problem, objective function and optimization constraints are significant components in the problem construction. The objective function and constraints of the cooperative optimization studied in this paper are formulated as follows.

4.1. Objective Function

The main optimization object is to improve the economics of the system allocation and decrease the cost of the microgrid operator. The multi-microgrid system discussed in the proposed model, which consists of several microgrids, is assumed to be owned by one single operator. Therefore, minimizing the total operation cost of the whole system is regarded as the objective function. The total cost can be described by Equations (4)–(7).

$$C_{MMS} = C_{RMG} + C_{OBMG} \tag{4}$$

$$C_{RMG} = \sum_{r=1}^{N_{RMG}} C_r^{RMG} \tag{5}$$

$$C_{OBMG} = \sum_{o=1}^{N_{OBMG}} C_o^{OBMG} \tag{6}$$

$$N_{MG} = N_{RMG} + N_{OBMG} \tag{7}$$

where C_{MMS} is total cost of the whole multi-microgrid system, C_{RMG} and C_{OBMG} are the total costs of RMGs and OBMGs, and C_r^{RMG} and C_o^{OBMG} are the costs of the rth RMG and the oth OBMG. N_{MG} is the number of all microgrids in the system. N_{RMG} and N_{OBMG} are numbers of RMGs and OBMGs.

The cost of each microgrid includes the costs of wind generation, solar generation, EVs and the cost of exchanging power with the main grid. It is calculated by Equation (8).

$$C_i^{MG} = C_i^{PV} + C_i^{WT} + C_i^{EV} + C_i^{G} \tag{8}$$

where C_i^{PV}, C_i^{WT}, C_i^{EV} and C_i^{EV} are costs of PV generation, wind power, EV operation, and energy exchanging with the main grid. Costs of PV generation and wind power are formulated by capital costs and costs of maintenance—they are constant values. Therefore, the costs of PV and wind generation can be calculated as Equation (9).

$$\begin{cases} C_i^{PV} = C_{PV} \cdot N_i^{PV} \\ C_i^{WT} = C_{WT} \cdot N_i^{WT} \end{cases} \tag{9}$$

where C_{PV} and C_{WT} are costs of single PV module and wind turbine, and N_i^{PV} and N_i^{WT} are numbers of PV modules and wind turbines in the ith microgrid.

Since the battery will depreciate with the increasing of charging/discharging cycles, extra charging and discharging caused by V2G services can significantly raise the cost of EV owners. In order to compensate for the EV owners, the extra cost caused by V2G should be considered. Considering depreciation expense, the cost the EVs can be calculated by Equation (10).

$$C_i^{EV} = N_i^{EV} \cdot C_{EVBCDI} + \sum_{t=1}^{T} \sum_{j=1}^{N_i^{EV}} P_{i,j}^{EV}(t) \cdot p^{EV}(t) + \sum_{j=1}^{N_i^{EV}} C_{depi,j}^{EV} \tag{10}$$

where N_i^{EV} is the number of EVs in the ith microgrid, T is the optimization period, C_{EVBCDI} is the cost of EVBCDI, P_t^{EV} is the price of EV charging and discharging, and C_{dep}^{EV} is the depreciation cost which is related to the charging/discharging cycles and is calculated according to [21].

Most of the energy in the main grid is generated from thermal plants, which produce large amount of pollutant as CO_2 and SO_2. In order to produce as little pollution as possible and to increase the utilization rate of RESs, energy exchanging with the main grid should be restricted. Therefore, in this paper, an environmental penalty is introduced. The costs of exchanging energy with the main grid can be calculated by Equation (11).

$$C_i^{G} = \sum_{t=1}^{T} p^{G}(t) \cdot P_i^{G}(t) + \sum_{t=1}^{T} C_i^{Gpen}(t) \tag{11}$$

where $p^{G}(t)$ is the prices of trading energy with the main grid, $P_i^{G}(t)$ is the exchanging power of the ith microgrid with the main grid in the tth hour, and $C_i^{Gpen}(t)$ is the environmental penalty of the tth hour.

$$C_i^{Gpen}(t) = [\max(0, P_i^{G}(t))] \cdot k_{env} \tag{12}$$

where k_{env} is the pollution factor, and it is determined by the types and proportion of pollutants [22].

4.2. Constraints

Some restrictions exist in the real system due to the characteristic of power devices and system balance. The constraints of the optimization problem in this paper are listed as follows.

4.2.1. Balance of Power Supply and Demand

The power generated and supplied by the microgrid should meet the power demand of end users in every dispatching time interval.

$$P_i^{EV}(t) + P_i^{WT}(t) + P_i^{PV}(t) + P_i^{G}(t) = P_i^{L}(t) \tag{13}$$

where $P_i^{PV}(t)$, $P_i^{WT}(t)$, and $P_i^{L}(t)$ are power of PV generation, wind generation, and load demand respectively.

4.2.2. Capacity Constraint of EV Batteries

Overcharge and discharge influence the life span of the batteries. However, in order to guarantee the driving demand of EV owners, SOC of EVs should be always higher than a certain minimum value.

$$SOC_{min}^{EV} \leq SOC_{i,j}^{EV}(t) \leq SOC_{max}^{EV} \tag{14}$$

4.2.3. Power Limits

Due to the physical limitations of EVBCDIs and connection transformers, both charging/discharging power and exchanging power are limited to a certain range.

$$\begin{cases} P_{min}^{EV} \leq P_{i,j}^{EV} \leq P_{max}^{EV} \\ P_{min}^{G} \leq P_{i,j}^{G} \leq P_{max}^{G} \end{cases} \tag{15}$$

4.2.4. Installation Constraints

To satisfy the charging demand of all EV users in the microgrid system, a total number of EVBCDIs in RMGs and the number of that in OBMGs should be equal, as demonstrated in (16) and (17).

$$\sum_{i=1}^{N_{MG}} N_i^{EV} = \sum_{r=1}^{N_{RMG}} N_r^{EVR} = \sum_{o=1}^{N_{OBMG}} N_o^{EVO} \tag{16}$$

$$N_{MG} = N_{RMG} + N_{OBMG} \tag{17}$$

where N_{RMG} and N_{OBMG} are numbers of RMGs and OBMGs, and N_r^{EVR} and N_o^{EVO} are the numbers of EVBCDIs in the rth RMG and the oth OBMG.

5. Two-Loop Optimization

5.1. Improved Particle Swarm Optimization

Since the optimization problem is a non-linear problem with a complex formulation of objective function, classic mathematical programming is not a proper method to solve it. Instead, an improved particle swarm optimization algorithm (IPSO) is used in this paper. In IPSO, each potential solution is regarded as a particle, and during the optimization, the particles update their positions and velocities in each loop iteration. After several rounds of iteration, the global best value of all the particles is adopted as the optimal solution.

As a kind of intelligent algorithm, IPSO can efficiently search for the optimal solution of a complex nonlinear optimization problem. Details of the algorithm can be seen in [23,24].

5.2. Two-Loop Optimization Process

The cooperative optimization has a two-loop structure. The process is demonstrated as follows in Figure 4.

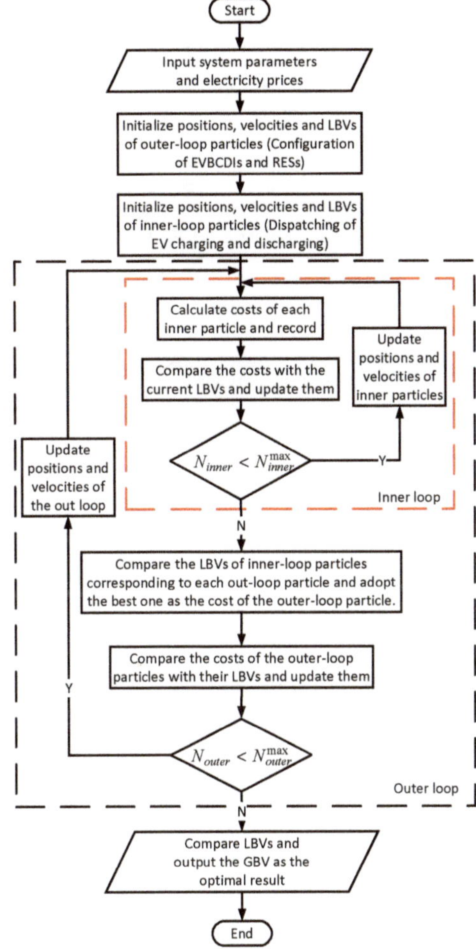

Figure 4. Process of the cooperative optimization. LBV: local best value, GBV: global best value.

As shown in the figure, the process consists of two loops, which correspond to optimizations of dispatching and system allocation.

The inner-loop is optimization of system dispatching. In this process, the system operation is optimized to realize the lowest total cost while the allocation of the system is fixed.

In the outer-loop, the allocation of EVs and RESs is optimized. By comparing the optimal total cost of different allocation cases, the allocation with the lowest total cost is adopted as an optimal allocation.

By employing the cooperative optimization method, economic benefits can be realized, and the system costs can be decreased.

6. Case Study and Discussion

6.1. Case Description

6.1.1. Case 1

In this case, neither numbers of EV charging/discharging infrastructures nor installation capacities of renewable energy sources are optimized. EVs will be discharged within the limits of SOCs in OBMG and be charged as much as possible when they return to RMGs. Economic dispatching of a multi-microgrid system is optimized by general management system (GMS). Case 1 is adopted as a reference in comparison to other cases.

6.1.2. Case 2

In this case, the numbers of EV charging/discharging infrastructures in different microgrids are optimized while the installation capacities of RESs are fixed. Similarly, economic dispatching of multi-microgrid system is optimized by GMS.

6.1.3. Case 3

In this case, the numbers of EV charging/discharging infrastructures are fixed while the installation capacities of RESs are optimized. Similarly, economic dispatching of multi-microgrid system is optimized by GMS.

6.1.4. Case 4

In this case, economic dispatching and capacity allocation of RESs and EVs in multi-microgrid system are optimized synergistically. EVs and RESs can cooperate best in this case, and there is no redundant installation of EVBCDIs or RESs.

6.2. Simulation System Construction

For simplification of the simulation, a multi-microgrid system with two RMGs and one OBMG is constructed. All EV owners are living in the two RMGs, identified as RMG1 and RMG2, and working in the OBMG. Exchanging power limit of the RMGs is $-95 \text{ kW} \leq P_i^G(t) \leq 95 \text{ kW}$, and the limit for OBMG is $-250 \text{ kW} \leq P_i^G(t) \leq 250 \text{ kW}$. Since the three microgrids are geographically near, PV generation and wind generation curves are similar and are forecasted by artificial neural networks using historical weather data derived from local weather records. Load demands of the three microgrids are derived from data of real microgrids with similar sizes.

A time-of-use electricity price is adopted as the energy exchanging prices with the main grid. The prices are demonstrated in Figure 5 [23]. In order to guarantee the benefits of the EV owners, charging and discharging prices of EVs should be restricted, as in (18)

$$p_C^{EV}(t) \leq \eta_C \cdot \eta_D \cdot p_D^{EV}(t) \tag{18}$$

where $p_C^{EV}(t)$ and $p_D^{EV}(t)$ are charging and discharging prices of EVs. They are assumed to be time-invariable in this paper. Additionally, to encourage participation of microgrids, discharging prices of EVs should be lower than the highest exchanging price in OBMG, and charging prices should be higher than the lowest exchanging price in RMG.

$$\begin{cases} p_C^{EV}(t) \geq \min(p_{RMG}^G(t)) \\ p_D^{EV}(t) \leq \max(p_{OBMG}^G(t)) \end{cases} \tag{19}$$

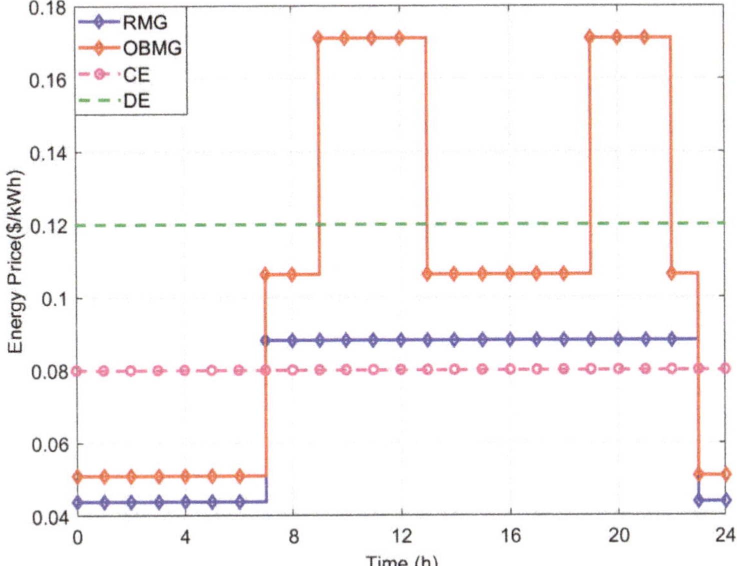

Figure 5. Prices of energy exchanging and EV charging and discharging.

For the EV model, BYD E6 is selected in the simulation. The capacity of the EV battery is 80 kWh and charging and discharging power limits are both 7 kW in normal charging mode. Charging and discharging efficiency are also identical, at 0.9. The daily cost of installing an EVBCDI is 0.45$. Some other data are referred to in [25].

6.3. Simulation Results

6.3.1. Case 1

In case one, the numbers of EVBCDIs and installation capacities of RESs are set to a fixed number according to the load demand and neither of them are optimized. The numbers of EVCDIs for RMG1, RMG2 and OBMG are 10, 10, and 20, respectively. WT installation capacities for RMG1, RMG2, and OBMG are 240 kW, 160 kW, and 320 kW. PV installation capacities for RMG1, RMG2, and OBMG are 180 kW, 120 kW, and 240 kW. The two-day operation profiles of RMG1 and OBMG in case one are shown in Figure 6. Profiles of RMG2 are similar to RMG1 and are not demonstrated here.

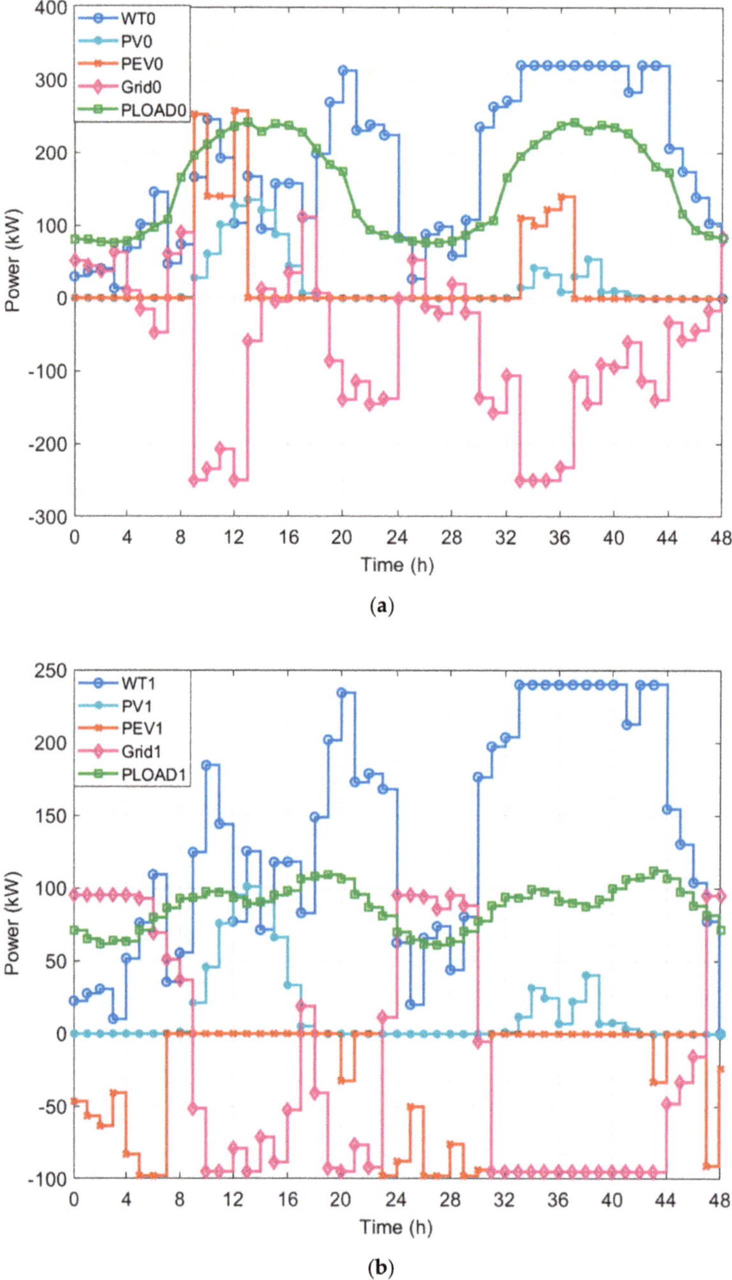

Figure 6. Dispatching results in Case 1: (**a**) Dispatching profiles of RMG1; (**b**) Dispatching profiles of OBMG.

As demonstrated in Figure 6, the EVs tend to charge in RMGs to satisfy their driving demands for the next day, and discharge in the OBMG to participate in the operation of the microgrid and earn some benefit by trading energy with the microgrid operator.

For most systems, the electricity prices of the main grid are always related to the load demand. The electricity prices are high in load peak hours, and the prices are low in load valley hours [26]. However, as shown in Figure 6, the microgrids tend to sell energy to the main grid when the electricity prices are high, and purchase energy from the main grid the prices of electricity are low. Therefore, the energy trading among microgrids and the grid can contribute to peak shaving of the grid.

In Case 1, the total cost of one year is $195,195.80. The costs for RMG1, RMG2, and OBMG are $67,872.80, $38,063.00, and $89,260.00, respectively.

6.3.2. Case 2

In Case 2, the numbers of EVBCDIs in three microgrids are optimized, while installation capacities of RESs are kept the same as Case 1. The optimal numbers of EVCDIs for RMG1, RMG2, and OBMG are 14, 15, and 29. WTs installation capacities for RMG1, RMG2, and OBMG are 240 kW, 160 kW, and 320 kW. PV installation capacities for RMG1, RMG2, and OBMG are 180 kW, 120 kW, and 240 kW. Since the two-day operation profiles of RMG1 are identical to those in Case 1, the operation profiles of RMG2 are shown here. The two-day operation profiles of RMG2 and OBMG in Case 2 are shown in Figure 7.

Tendency of EV charging and discharging is similar with Case 1. However, the numbers of EVCDIs in both RMGs and OBMG are optimized. Comparing Figure 7 with Figure 6, it is obvious that the total charging and discharging power of EVs increased. The total energy transmitted between RMGs and OBMG increased, and more energy with lower prices are transported to the places and times with higher prices. Therefore, the operation cost of the system decreased, and the total cost also decreased.

In Case 2, total cost of one year is $193,867.50. The costs for RMG1, RMG2, and OBMG are $67,491.10, $37,374.50, and $89,001.90, respectively.

6.3.3. Case 3

In Case 3, the numbers of EVBCDIs in three microgrids are fixed to be the same with Case 1, and installation capacities of RESs are optimized. The numbers of EVCDIs for RMG1, RMG2, and OBMG are 10, 10, and 20 respectively. Optimal WT installation capacities for RMG1, RMG2, and OBMG are 80 kW, 40 kW, and 360 kW. The installation capacity of PV modules for RMG1, RMG2, and OBMG are 180 kW, 0 kW, and 400 kW. The 2-day operation profiles of RMG1 and OBMG in Case 3 are shown in Figure 8.

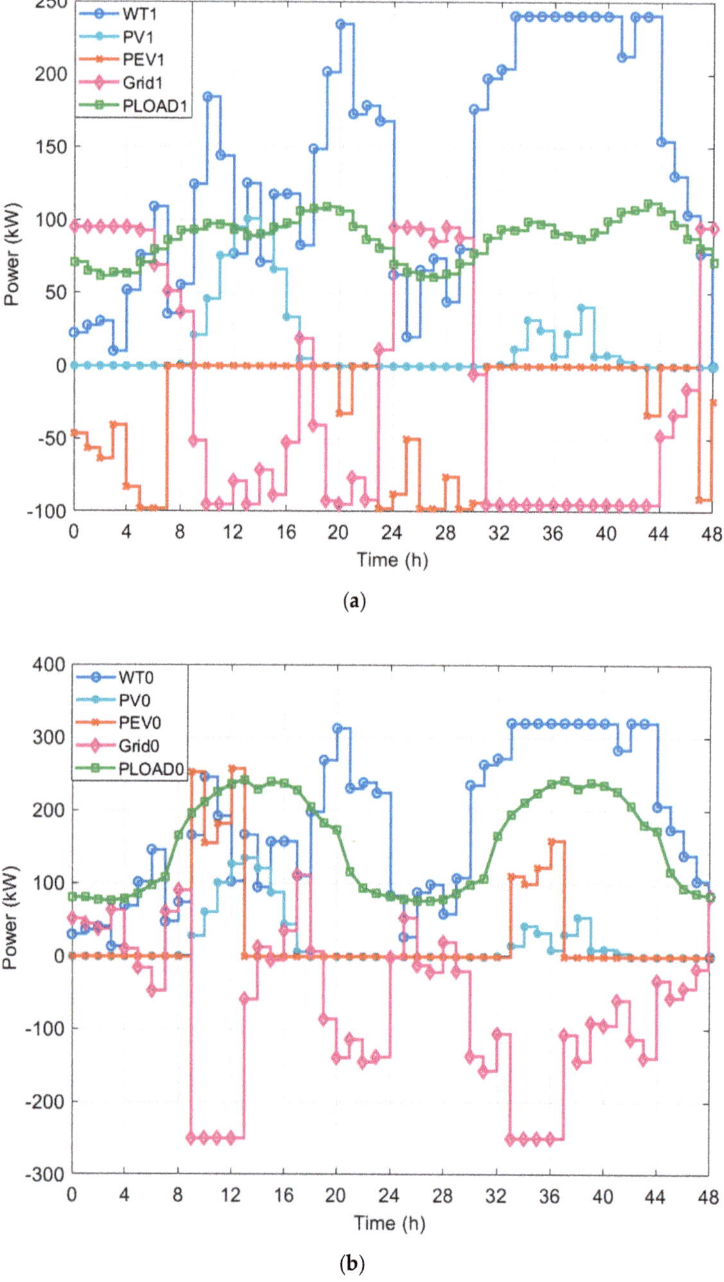

Figure 7. Dispatching results in Case 2: (**a**) Dispatching profiles of RMG1; (**b**) Dispatching profiles of OBMG.

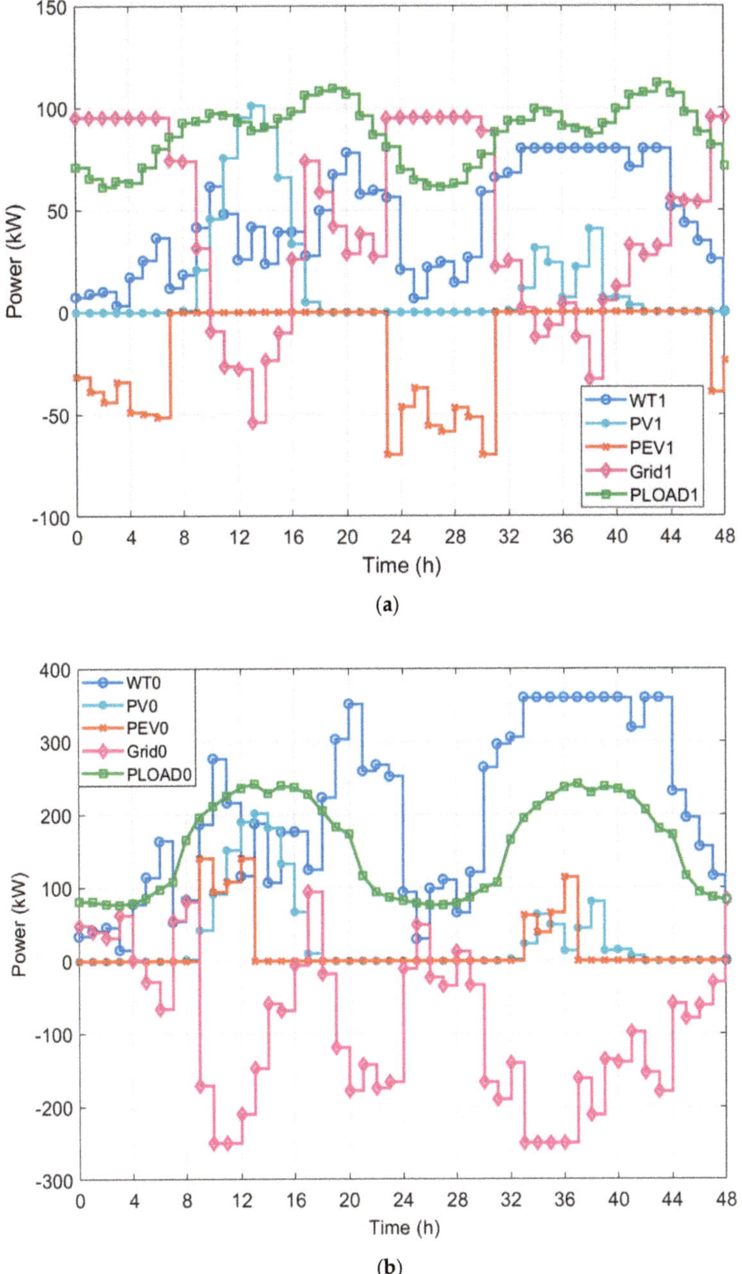

Figure 8. Dispatching results in Case 3: (**a**) Dispatching profiles of RMG1; (**b**) Dispatching profiles of OBMG.

According to the figures, we can see that the profiles of RMG1 are not much different. The installation capacities of RESs in RMGs and OBMG have been optimized. For the RMGs, electricity prices of the distribution grid are low and it is not beneficial to sell energy to the grid. Therefore, the redundant installation of RESs is decreased, and less energy are sold to the grid while the load demands of RMGs are still satisfied. For OBMG, the electricity prices of the distribution grid are high, and it is beneficial to sell extra energy to the grid. Therefore, more RESs are installed to sell more energy to the grid, and the microgrid earns more from selling energy.

In Case 3, the total cost of one year is $161,090.30. The costs for RMG1, RMG2, and OBMG are $55,138.10, $36,348.80, and $69,603.40, respectively.

6.3.4. Case 4

In Case 4, both the numbers of EVBCDIs of the three microgrids and installation capacities of RESs are optimized. The numbers of EVCDIs for RMG1, RMG2, and OBMG are 19, 8, and 27, respectively. WTs installation capacities for RMG1, RMG2, and OBMG are 80 kW, 80 kW, and 360 kW. PV installation capacities for RMG1, RMG2, and OBMG are 120 kW, 0 kW, and 400 kW. The two-day operation profiles of RMG1 and OBMG in Case 1 are shown in Figure 9.

In this case, both EV charging devices and RESs are optimally allocated. Since the cost of PV is higher than wind generation and since the outputs of PV modules are unstable, the installation of PV modules is decreased. Meanwhile, the number of EVCDIs and the installation capacities of RESs are optimally matched. Neither redundant RES installation nor redundant installation of EVCDIs exists. Therefore, the total cost in this case is the lowest.

In Case 4, the total cost of one year is $159,264.80. The costs for RMG1, RMG2, and OBMG are $56,012.30, $36,306.40, and $66,946.10, respectively.

6.4. Comparison

Optimization results of system allocation are demonstrated in Table 1. Results of costs and the net exchanged energy between microgrids and the main grid in four cases are demonstrated in Table 2. The total costs and costs of the three microgrids in the four cases are listed. In addition, the costs of allocation and dispatching are also calculated and listed in the table. According to the figures in Section 6.3 and the data in the tables, conclusions can be drawn as follows:

1. After optimization, the net exchanged energy between microgrids and the main grid decreases, which means that the microgrids becomes more independent from the main grid.
2. Microgrids purchase energy from the main grid in load valley hours when electricity prices are low and sell energy to the main grid in load peak hours when prices are high, which contributes to the peak shaving of the main grid.
3. In Cases 3 and 4, numbers of PVs in RMGs dramatically decrease, while number of PVs in OBMG is greatly increased. The reason is that for RMGs, the EV charging load is mainly distributed during night and the output of PVs at night is zero. Therefore, PV modules in RMGs are reduced to save installation costs.
4. The total costs in Cases 2 and 3 are lower than that in Case 1. In Case 2, the numbers of EVBCDIs are optimized. Since redundant installation of EVBCDIs is cut down, the allocation is more proper and total cost is reduced. Similarly, in Case 3, the installation of RESs is optimized. The installation capacities of RESs are optimally configured according to the demand of the different kinds of microgrids, and the total cost is decreased compared to Case 1.
5. In Case 4, economic dispatching and capacity allocation of RESs and EVs in the multi-microgrid system are optimized synergistically. Though the allocation cost of RESs is slightly increased, the dispatching cost of the system is significantly reduced because of the cooperative optimization of economic dispatching and capacity allocation for RESs and EVs. The total cost in Case 4 is the lowest of all four cases. The integration efficiency of RESs and EVs is improved, and more environmental benefits are achieved.

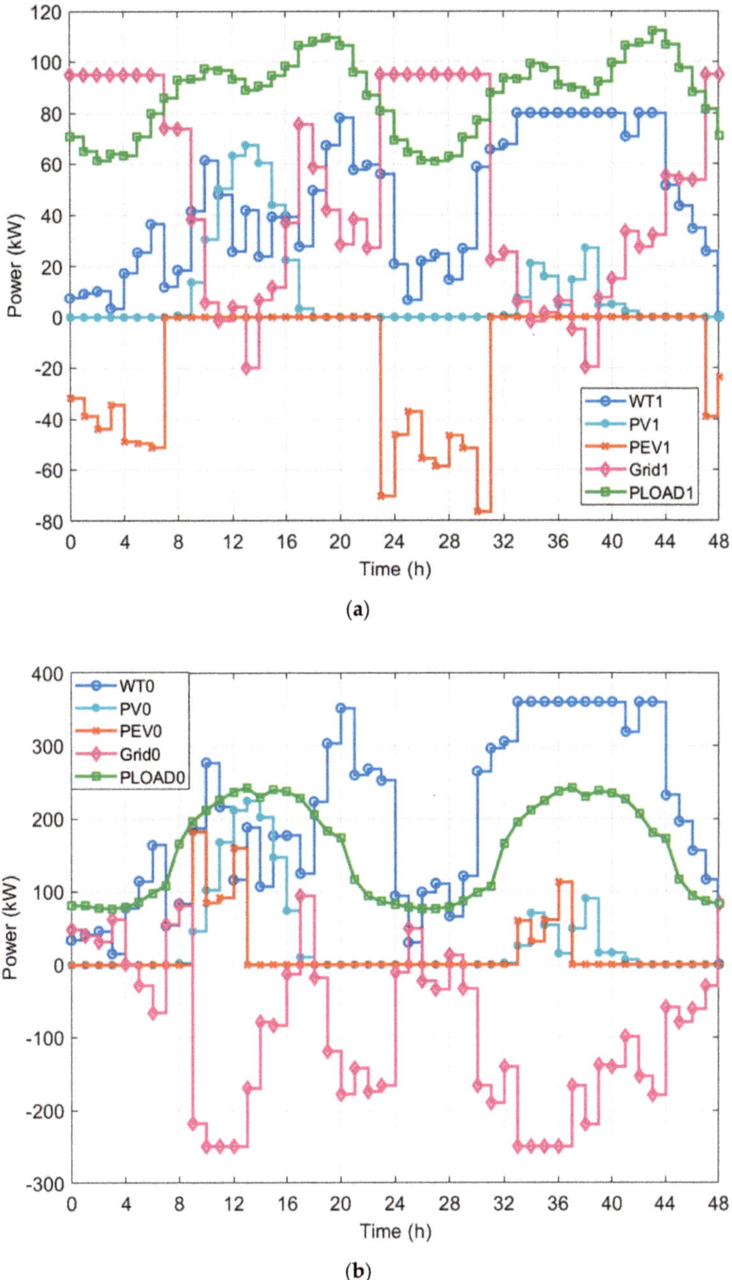

Figure 9. Dispatching results in Case 4: (**a**) Dispatching profiles of RMG1; (**b**) Dispatching profiles of OBMG.

Table 1. Comparison of Allocations in all Four Cases.

		Case 1	Case 2	Case 3	Case 4
EVs	RMG1	10	14	10	14
	RMG2	10	15	10	12
WTs	RMG1	6	6	2	2
	RMG2	4	4	1	2
	OBMG	8	8	9	9
PVs	RMG1	9	9	9	6
	RMG2	6	6	0	0
	OBMG	12	12	18	20
Allocation cost of EVs ($)		6570.0	9526.5	6570.0	8541.0
Allocation cost of RESs ($)		88,640.9	88,640.9	70,174.1	72,020.8
Total cost of allocation ($)		95,210.9	98,167.4	76,744.1	80,561.8

Table 2. Comparison of Costs and Net Exchanged Energy in all Four Cases.

	Case 1	Case 2	Case 3	Case 4
Cost of RMG1 ($)	67,872.8	67,491.1	55,053.9	56,012.3
Cost of RMG2 ($)	38,063.0	37,374.5	36,345.5	36,306.4
Cost of OBMG ($)	89,260.0	89,001.9	69,793.4	66,946.1
Net Exchanged Energy (MWh)	413.32	377.10	255.30	189.30
Dispatching cost ($)	97,028.4	95,700.1	84,448.7	78,703.0
Total Cost ($)	195,195.8	193,867.5	161,192.8	159,264.8

7. Conclusions

In this paper, a cooperative optimization method for a multi-microgrid system was proposed. Mathematical models of multi-microgrid systems including RESs and EVs were constructed. Energy cooperation of microgrids in the multi-microgrid system was accomplished via the mobile energy storage of EVs. The impact of ATSET of EVs on economic dispatching and capacity allocation of a multi-microgrid system was analyzed. A two-loop optimization methodology that considered the cooperation of economic dispatching and system allocation was proposed, and an IPSO algorithm was used to solve the optimization problem. Both EVs and RESs were succesfully optimized in the optimization. Four cases were presented to verify the cooperative optimization. By comparing the results of the four cases, the following conclusions are drawn:

1. By ATSET of EVs, energy can be transmitted through different microgrids, and economic dispatching of microgrids in the regional multi-microgrid system is cooperatively optimized. Optimized energy exchanging between regional multi-microgrid system and the main grid contributes to the peak shaving of the main grid.
2. By optimizing both economic dispatching and capacity allocation of RESs and EVs, the total cost of the regional multi-microgrid system is dramatically reduced, and economic benefit is achieved. After a cooperative optimization, the independency of the microgrids' operation is raised and integration efficiency of RESs and EVs is improved.

It is worth noting that the simulation system discussed in this paper was significantly simplified in order to clearly demonstrate the main methodology of the paper. Future works may focus on the application of the model to more complex systems. Real systems are always more complex, with various kinds of generation and load units, and some realistic factors such as the influence of the battery's remnant capacity on the behavior of EV owners can also be considered. A real system may also include storage batteries and micro-turbines. The way of applying the proposed model on real systems is an important topic of the future works. Moreover, since the charging and discharging prices of EVs considered in this paper are constant, a more flexible pricing system for EVs' charging and discharging is also an issue that worth studying.

Author Contributions: Conceptualization, J.C. and C.C.; methodology, J.C. and C.C.; software, J.C.; validation, J.C.; formal analysis, J.C. and C.C.; investigation, J.C. and C.C.; resources, J.C. and C.C.; data curation, J.C. and C.C.; writing—original draft preparation, J.C.; writing—review and editing, J.C. and C.C.; visualization, J.C.; supervision, C.C. and S.D.; project administration, C.C. and S.D.; funding acquisition, C.C.

Funding: This research was funded by the National Natural Science Foundation of China, grant number 51477067 and the Lite-On Power Electronics Technology Research Fund, grant number PRC20161047.

Conflicts of Interest: The authors declare no conflict of interest.

References

1. Wang, X.; Barnett, A. The evolving value of photovoltaic module efficiency. *Appl. Sci.* **2019**, *9*, 1227. [CrossRef]
2. Mohammad, A.M.; Ahmad, S.Y.; Behnam, M.I.; Mousa, M.; Miadreza, S.; João, P.S.C. Stochastic network-constrained co-optimization of energy and reserve products in renewable energy integrated power and gas networks with energy storage system. *J. Clean. Prod.* **2019**, *223*, 747–758.
3. Mei, F.; Wu, Q.; Shi, T.; Lu, J.; Pan, Y.; Zheng, J. An ultrashort-term net load forecasting model based on phase space reconstruction and deep neural network. *Appl. Sci.* **2019**, *9*, 1487. [CrossRef]
4. Ross, M.; Hidalgo, R.; Abbey, C.; Joós, G. Energy storage system scheduling for an isolated microgrid. *IET Renew. Power Gener.* **2011**, *5*, 117–126. [CrossRef]
5. Zhang, M.; Chen, J. The energy management and optimized operation of electric vehicles based on microgrid. *IEEE Trans. Power Deliv.* **2014**, *29*, 1427–1435. [CrossRef]
6. Van Der Meer, D.; Mouli, G.R.C.; Mouli, G.M.E.; Elizondo, L.R.; Bauer, P. Energy management system with PV power forecast to optimally charge EVs at the workplace. *IEEE Trans. Ind. Inform.* **2018**, *14*, 311–320. [CrossRef]
7. Meyer, D.; Wang, J. Integrating ultra-fast charging stations within the power grids of smart cities: A review. *IET Smart Grid* **2018**, *1*, 3–10. [CrossRef]
8. Morais, H.; Sousa, T.; Vale, Z.; Faria, P. Evaluation of the electric vehicle impact in the power demand curve in a smart grid environment. *Energy Convers. Manag.* **2014**, *82*, 268–282. [CrossRef]
9. Hu, J.; Morais, H.; Sousa, T.; Lind, M. Electric vehicle fleet management in smart grids: A review of services, optimization and control aspects. *Renew. Sustain. Energy Rev.* **2016**, *56*, 1207–1226. [CrossRef]
10. Cerna, F.V.; Pourakbari-Kasmaei, M.; Contreras, J.; Gallego, L.A. Optimal selection of navigation modes of HEVs considering CO2 emissions reduction. *IEEE Trans. Veh. Technol.* **2019**, *68*, 2196–2206. [CrossRef]
11. Parinaz, A.; Behnam, M.I.; Manijeh, A.; Mehdi, A.; Kazem, Z. Optimal scheduling of plug-in electric vehicles and renewable micro-grid in energy and reserve markets considering demand response program. *Energies* **2018**, *186*, 293–303.
12. Zhang, H.; Hu, Z.; Xu, Z.; Song, Y. An integrated planning framework for different types of an integrated planning framework for different types of PEV charging facilities in urban area. *IEEE Trans. Smart Grid* **2014**, *7*, 2273–2284. [CrossRef]
13. Ugirumurera, J.; Haas, Z. Optimal sizing of a completely green charging system for electric vehicles. *IEEE Trans. Transp. Electrif.* **2017**, *3*, 565–577. [CrossRef]
14. Yang, H.; Pan, H.; Luo, F.; Qiu, J.; Deng, Y.; Lai, M.; Zhao, Y. Operational Planning of Electric Vehicles for Balancing Wind Power and Load Fluctuations in a Microgrid. *IEEE Trans. Sustain. Energy* **2017**, *8*, 592–604. [CrossRef]
15. Lin, Y.; Dong, P.; Sun, X.; Liu, M. Two-level game algorithm for multi-microgrid in electricity market. *IET Renew. Power Gener.* **2017**, *11*, 1733–1740. [CrossRef]
16. Aluisio, B.; Conserva, A.; Dicorato, M.; Forte, G.; Trovato, M. Optimal operation planning of V2G-equipped Microgrid in the presence of EV aggregator. *Electr. Power Syst. Res.* **2017**, *152*, 295–305. [CrossRef]
17. Sideratos, G.; Hatziargyriou, N.D. An advanced statistical method for wind power forecasting. *IEEE Trans. Power Syst.* **2007**, *22*, 258–265. [CrossRef]
18. Chen, C.; Duan, S.; Cai, T.; Liu, B. Online 24-h solar power forecasting based on weather type classification using artificial neural network. *Sol. Energy* **2011**, *85*, 2856–2870. [CrossRef]

19. Chen, C.; Duan, S. Optimal integration of plug-in hybrid electric vehicles in microgrids. *IEEE Trans. Ind. Inform.* **2014**, *10*, 1917–1926. [CrossRef]

20. Hu, J.; Si, C.; Lind, M.; Yu, R. Preventing distribution grid congestion by integrating indirect control in a hierarchical electric vehicles' management system. *IEEE Trans. Transp. Electrif.* **2016**, *2*, 290–299. [CrossRef]

21. Ortega-Vazquez, M. Optimal scheduling of electric vehicle charging and vehicle-to-grid services at household level including battery degradation and price uncertainty. *IET Gener. Transm. Distrib.* **2014**, *8*, 1007–1016. [CrossRef]

22. Qian, K.; Zhou, C.; Yuan, Y. Analysis of the environmental benefits of distributed generation. In Proceedings of the 2008 IEEE Power and Energy Society General Meeting-Conversion and Delivery of Electrical Energy in the 21st Century, Pittsburgh, PA, USA, 20–24 July 2008; pp. 1–5.

23. Coello Coello, C.A.; Pulido, G.T.; Lechuga, M.S. Handling multiple objectives with particle swarm optimization. *IEEE Trans. Evol. Comput.* **2004**, *8*, 256–279. [CrossRef]

24. Saber, A.Y.; Venayagamoorthy, G.K. Plug-in vehicles and renewable energy sources for cost and emission reductions. *IEEE Trans. Ind. Electron.* **2011**, *58*, 1229–1238. [CrossRef]

25. Chen, C.; Xiao, L.; Duan, S.; Chen, J. Cooperative optimization of electric vehicles in microgrids considering across-time-and-space energy transmission. *IEEE Trans. Ind. Electron.* **2019**, *66*, 1532–1542. [CrossRef]

26. Ghosh, A.; Aggarwal, V. Control of charging of electric vehicles through menu-based pricing. *IEEE Trans. Smart Grid* **2018**, *9*, 5918–5929. [CrossRef]

Article

Adaptive Dual Extended Kalman Filter Based on Variational Bayesian Approximation for Joint Estimation of Lithium-Ion Battery State of Charge and Model Parameters

Jing Hou [1,†], Yan Yang [1,*], He He [2] and Tian Gao [1]

[1] School of Electronic and Information, Northwestern Polytechnical University, Xi'an 710072, China; jhou0825@nwpu.edu.cn (J.H.); tiangao@nwpu.edu.cn (T.G.)
[2] System Engineering Research Institute of CSSC, Beijing 100094, China; hehe5951@163.com
* Correspondence: yangyan7003@nwpu.edu.cn
† Current address: No. 127, Youyi West Road, Xi'an 710072, China.

Received: 4 April 2019; Accepted: 23 April 2019; Published: 26 April 2019

Abstract: An accurate state of charge (SOC) estimation is vital for the safe operation and efficient management of lithium-ion batteries. At present, the extended Kalman filter (EKF) can accurately estimate the SOC under the condition of a precise battery model and deterministic noise statistics. However, in practical applications, the battery characteristics change with different operating conditions and the measurement noise statistics may vary with time, resulting in nonoptimal and even unreliable estimation of SOC by EKF. To improve the SOC estimation accuracy under uncertain measurement noise statistics, a variational Bayesian approximation-based adaptive dual extended Kalman filter (VB-ADEKF) is proposed in this paper. The variational Bayesian inference is integrated with the dual EKF (DEKF) to jointly estimate the lithium-ion battery parameters and SOC. Meanwhile, the measurement noise variances are simultaneously estimated in the SOC estimation process to compensate for the model uncertainties, so that the adaptability of the proposed algorithm to dynamic changes in battery characteristics is greatly improved. A constant current discharge test, a pulse current discharge test, and an urban dynamometer driving schedule (UDDS) test are performed to verify the effectiveness and superiority of the proposed algorithm by comparison with the DEKF algorithm. The experimental results show that the proposed VB-ADEKF algorithm outperforms the traditional DEKF algorithm in terms of SOC estimation accuracy, convergence rate, and robustness.

Keywords: state of charge (SOC); joint estimation; lithium-ion battery; variational Bayesian approximation; dual extended Kalman filter (DEKF); measurement statistic uncertainty

---

1. Introduction

Electric vehicles (EVs) are believed worldwide to be one of the most important development directions in the vehicle industry because of their advantages in low pollution and energy saving. Lithium-ion batteries, by virtue of their high energy and power density, are the fundamental power source of EVs [1]. Nevertheless, lithium-ion batteries have hidden dangers in safety. Once over-charged or over-discharged, the battery capacity will drop, leading to reduced lifetime or damage, or even explosion. Moreover, the battery characteristics change with different operating conditions. Therefore, to ensure the safety, reliability, and efficiency of the battery system, a battery management system (BMS) is developed and utilised. The BMS is used to control and monitor the battery operating conditions to guarantee the safe operation and longevity of the battery.

The state of charge (SOC) is a significant parameter which indicates the amount of remaining energy in the battery for further service. It needs to be estimated accurately online in order to prevent any over-charge and over-discharge, while providing the information and support for the effective and flexible operation of the vehicle. However, it is difficult to acquire precise SOC estimates since the battery itself is a highly nonlinear system and has a lot of uncertainties.

Extensive studies have been made into SOC estimation. Hannan et al. [2] gave a comprehensive review of these research fruits. Overall, these methods can be mainly classified into four categories: (1) the open-circuit voltage (OCV) method; (2) current integration or coulomb counting (CC) method; (3) data-driven methods; and (4) model-based methods.

The OCV method is based on the one-to-one corresponding relationship of open-circuit voltage and SOC. It is simple but susceptible to temperature, battery aging, and other factors, and the acquisition of accurate measurement of OCV needs lots of rest time, making it almost impossible for moving vehicles. The CC method is widely used in many applications, but it has two weak points. First, its estimation accuracy is strongly dependent on the correctness of the initial SOC value, and second, it can easily diverge due to error accumulation in the current measurement.

The data-driven methods include neural network (NN) algorithms [3–5], fuzzy logic (FL) algorithms [6–8], support vector machines (SVMs) [9–11], and so on. For instance, He et al. [3] employed a multilayer feedforward neural network to estimate the SOC of Lithium-ion batteries by use of the discharge current, terminal voltage, and temperature of the battery as input. Zhao et al. [4] combined a back propagation neural network and adaptive Kalman filter to estimate the SOC, and used a forgetting-factor recursive least-squares (FFRLS) algorithm to identify the time-varying battery model parameter. In [6], a fuzzy logic algorithm was applied to estimate SOC of lithium-ion batteries for the application of a portable defibrillator. AC impedance and voltage recovery measurements were used as the input parameters for the FL model. In [7], a more advanced algorithm named adaptive neuro-fuzzy inference system (ANFIS) was developed to estimate SOC. In addition, Hu et al. [9] proposed an SOC estimation based on an optimized support vector machine for regression with double search optimization process. In [10], an SOC estimation method based on fuzzy least-squares SVM was proposed. These data-driven algorithms do not need to know any battery characteristics and have a good ability of nonlinear mapping and self-learning. However, they require a large amount of experimental data to train the intelligent model beforehand. Different types of training data and training methods exert a great influence on the model error. When the training data cannot cover all the operating conditions, for example, the experiments are incomplete or the battery characteristics have changed, the SOC estimation accuracy will decrease.

The model-based methods are dominated by Kalman filter and its derivatives. Chen et al. [12] used an extended Kalman filter (EKF) along with a nonlinear battery model to estimate the SOC of the lithium-ion battery. In order to further improve the accuracy, adaptive extended Kalman filter (AEKF), unscented Kalman filter (UKF), and adaptive unscented Kalman filter (AUKF) are proposed. He et al. [13] identified the parameters of an improved Thevenin battery model using EKF, and then adopted an AEKF for obtaining correct and robust SOC of the lithium-ion batteries. In [14], a UKF-based method was used to self-adjust the model parameters and provide state of charge estimation of the battery. References [15,16] adopted AUKF to realize SOC estimation. In addition, an adaptive cubature Kalman filter (CKF) was proposed in [17] to improve the convergence rate and SOC estimation accuracy. To overcome the accuracy degradation caused by non-Gaussian noise, particle filters were utilized to estimate the SOC of batteries in [18,19]. To compensate for the time-variability of battery parameters due to variational operating conditions and battery aging, a dual EKF and a dual UKF were employed in [20,21] for simultaneous SOC and parameters estimation, respectively. The model-based methods eliminate the need for correct initial SOC values and accurate measurement, which are requisite for the CC method, and have no demand for a large amount of train data, so it is widely applied and is studied extensively. However, the estimation accuracy is strongly affected by the battery model and parameters. Once there is a model mismatch, the estimation performance will rapidly decline.

In practical applications, the battery model parameters will change with SOC, temperature, and the degree of battery aging. Moreover, the statistic information of the process noise and sensor noise may be unknown or time-varying. Under these situations, the traditional KF-based methods will have low estimation accuracy and poor robustness. To address this issue, some researchers [22–24] resort to a H∞ filter, which takes the time-varying battery parameter into account and has no need to know the statistics of the process noise and the measurement noise. It has strong robustness under uncertain conditions. However, the H∞ filter is a tradeoff between robustness and accuracy, so the SOC estimation accuracy is sacrificed to some extent for the robustness. Liu et al. [25] combined the idea of square root filter and adaptive unscented Kalman filter (ASRUKF) algorithm based on improved Sage-Husa estimation to estimate the SOC of a lithium-ion battery. This method adaptively adjusted the values of the process and measurement covariances in the estimation process to improve the accuracy of SOC estimation. However, it does not consider the uncertainties brought by varying battery model parameters. EI Din et al. [26] proposed a multiple-model EKF (MM-EKF) and an autocovariance least squares (ALS) method for estimating the SOC of lithium-ion battery cells. MM-EKF reduced the dependence of the EKF algorithm on the correct assumptions of the measurement noise statistics by weighted summation of the estimates of multiple hypothesized EKFs. The ALS method extracted the possible correlation in the innovation sequence to estimate the measurement noise covariance. However, both methods leave out consideration of the time-varying battery parameters. Furthermore, the computation load is larger compared with the conventional EKF algorithm.

In fact, the Bayesian approach is the most general approach of solving the problem with uncertain parameters. But it is hard to get the analytical solution for most Bayesian approaches due to complex probability density function or high dimension of integration. Recently, the variational Bayesian (VB) inference method [27–31] has drawn extensive attentions, which utilizes a new simpler, analytically tractable distribution to approximate the true posterior distribution so as to avoid the direct complex calculation of multi-dimensional probability density function. Sarkka et al. [27] adopted the VB method for joint recursive estimation of the dynamic state and the time-varying measurement noise parameters in linear state space models. Li et al. [28] employed VB approximation for the unscented Kalman filter to estimate the time-varying measurement noise covariance so as to improve algorithm adaptability. Sun et al. [29] proposed a VB method to estimate the system states with unknown inputs. Hou et al. [30] combined the VB method with the shifted Rayleigh filter to jointly estimate the target position and the clutter probability so that improving the performance of bearings-only target tracking. In [31], the VB approach was applied to estimate the ARX model parameters along with time delays.

In this paper, we combine the idea of variational Bayesian inference with the dual EKF algorithm (VB-ADEKF) to jointly estimate the battery parameters and SOC of lithium-ion batteries of electric vehicles. Meanwhile, the measurement noise variances are simultaneously estimated with the state estimation to account for the battery model uncertainties and measurement uncertainties, minimizing the impact of model mismatch. The effectiveness of the proposed algorithm has been verified through experiments under different operating conditions. The results show that the proposed VB-ADEKF algorithm outperforms the dual EKF algorithm in terms of SOC estimation accuracy, convergence rate, and robustness. The mean SOC estimation error of VB-ADEKF is under 1% for most of the time.

The remainder of this paper is organized as follows: Section 2 describes the battery model, the definition of SOC and establishes the state space models for SOC estimation and battery parameter estimation. Section 3 illustrates the variational Bayesian approximation-based adaptive Kalman filter and Section 4 presents the proposed variational Bayesian approximation-based adaptive dual extended Kalman filter. The experimental verification and analysis are presented in Section 5. Finally, Section 6 provides a conclusion.

2. Battery Modeling

2.1. Battery Model

For the accurate estimation of SOC, a reliable battery model is required. Considering the model accuracy, the structure complexity, and the computation time, the first order resistor–capacitor (RC) model as shown in Figure 1 is adopted to model the lithium-ion battery.

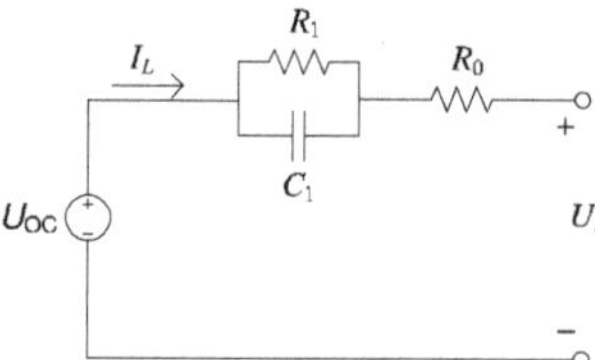

Figure 1. A first-order resistor–capacitor (RC) battery model.

The electrical behavior of the model can be written as follows:

$$U_t = U_{oc} - U_1 - I_L R_0 \tag{1}$$

$$\dot{U}_1 = \frac{I_L}{C_1} - \frac{U_1}{R_1 C_1} \tag{2}$$

where U_t denotes the terminal voltage of the battery, U_{oc} is the open-circuit voltage, U_1 is the polarization voltage of the RC network, I_L is the load current, R_0 represents the ohmic internal resistance, and R_1 and C_1 represent the polarization resistance and polarization capacitance, respectively.

The nonlinear relationship between the open-circuit voltage and the SOC is described using the fifth-order polynomial model as:

$$U_{oc}(SOC) = k_0 + k_1 SOC + k_2 SOC^2 + k_3 SOC^3 + k_4 SOC^4 + k_5 SOC^5 \tag{3}$$

where $k_0, k_1, k_2, k_3, k_4, k_5$ are the parameters to be identified.

2.2. Definition of State of Charge

The state of charge (SOC) is defined as the ratio of the remaining capacity in a battery over the rated battery capacity. Using the CC method, the battery SOC can be calculated as follows:

$$SOC_t = SOC_{t_0} - \frac{1}{Q_{rate}} \int_{t_0}^{t} \eta_c I_{L,t} dt \tag{4}$$

where η_c is the coulomb efficiency, $I_{L,t}$ is the load current at time t. Q_{rate} is the rated capacity of the battery.

2.3. State-Space Model

2.3.1. State Space Model for SOC Estimation

Taking $X = [SOC, U_1]^T$ as the state vector, the load current I_L as the input and the terminal voltage U_t as the output, we can obtain the discrete state-space model as

$$\begin{cases} X_{k+1} = f(X_k, I_{L,k}, \theta_k) + w_k \\ y_k = U_{t,k} = h(X_k, I_{L,k}, \theta_k) + v_k \end{cases} \tag{5}$$

where $\theta_k = [R_0, R_1, \tau_1]^{\mathrm{T}}$ represents the time-varying battery model parameter vector; $\tau_1 = R_1 C_1$ is the time constant of the RC network; $w_k \sim \mathcal{N}(0, Q_k^x)$ is the Gaussian process noise with covariance Q_k^x; $v_k \sim \mathcal{N}(0, \Sigma_k^x)$ is the measurement noise with variance Σ_k^x. The initial state has a Gaussian prior distribution $X_0 \sim \mathcal{N}(\hat{X}_0, P_0)$. The state prior and process noise are assumed to be known, while the measurement noise variance Σ_k^x is assumed to be unknown. In addition, the process noise, measurement noise, and initial state value are independent of each other.

$f(\cdot)$ and $h(\cdot)$ represent the nonlinear functions of state vector X_k, input $I_{L,k}$, and battery model parameter vector θ_k. Their mathematical expressions are

$$f(\cdot) = \begin{bmatrix} 1 & 0 \\ 0 & e^{-\frac{\Delta t}{\tau_1}} \end{bmatrix} \begin{bmatrix} SOC_k \\ U_{1,k} \end{bmatrix} + \begin{bmatrix} -\frac{\eta_c \Delta t}{Q_{rate}} & 0 \\ 0 & R_1(1 - e^{-\frac{\Delta t}{\tau_1}}) \end{bmatrix} I_{L,k} \tag{6}$$

$$h(\cdot) = U_{OC}(SOC_k) - U_{1,k} - I_{L,k} R_0 \tag{7}$$

where Δt is the sampling interval of the current.

The state transition matrix and the input control matrix are respectively

$$F_k = \begin{bmatrix} 1 & 0 \\ 0 & e^{-\frac{\Delta t}{\tau_1}} \end{bmatrix} \tag{8}$$

$$G_k = \begin{bmatrix} -\frac{\eta_c \Delta t}{Q_{rate}} & 0 \\ 0 & R_1(1 - e^{-\frac{\Delta t}{\tau_1}}) \end{bmatrix} \tag{9}$$

Linearizing the measurement function, we can get the Jacobian measurement matrix as

$$H_{x,k} = \left. \frac{\partial h(\cdot)}{\partial X_k} \right|_{X_k = \hat{X}_k^-} = \begin{bmatrix} \frac{\partial U_{OC}}{\partial SOC_k} & -1 \end{bmatrix} \tag{10}$$

2.3.2. State Space Model for Battery Parameter Estimation

Because the battery model parameters vary with changes in the batteries' SOC, degree of aging, and environmental temperature, online recursive battery parameter estimation is needed. So here we establish the state-space equations of the battery parameters as:

$$\begin{cases} \theta_{k+1} = \theta_k + r_k \\ d_k = h(X_k, I_{L,k}, \theta_k) + e_k \end{cases} \tag{11}$$

where $r_k \sim \mathcal{N}(0, Q_k^\theta)$ is a small white noise with covariance Q_k^θ that reflects the time-varying parts of the parameters, d_k is a measurement function of θ_k, and $e_k \sim \mathcal{N}(0, \Sigma_k^\theta)$ is the measurement noise to account for the sensor noise and modeling uncertainties. Σ_k^θ is assumed unknown here.

The Jacobian measurement matrix of θ_k is calculated as follows [20]:

$$H_{\theta,k} = \left. \frac{dh(\cdot)}{d\theta_k} \right|_{\theta_k = \hat{\theta}_k^-} = \frac{\partial h(\cdot)}{\partial \theta_k^-} + \frac{\partial h(\cdot)}{\partial \hat{X}_k^-} \cdot \frac{d\hat{X}_k^-}{d\theta_k^-} \tag{12}$$

$$\frac{\partial h(\cdot)}{\partial \theta_k^-} = \begin{bmatrix} -I_{L,k} & 0 & 0 \end{bmatrix} \tag{13}$$

$$\frac{\partial h(\cdot)}{\partial \hat{X}_k^-} = H_{x,k} \tag{14}$$

$$\frac{d\hat{X}_k^-}{d\hat{\theta}_k^-} = \begin{bmatrix} 0 & 0 & 0 \\ 0 & a_{22} & a_{23} \end{bmatrix} \tag{15}$$

where

$$a_{22} = -I_{L,k} \cdot (\exp(\Delta t / \tau_1^2) - 1)$$
$$a_{23} = (\Delta t / \tau_1^2) \cdot (\hat{x}_{2,k}^- - R_1 I_{L,k}) \cdot \exp(\Delta t / \tau_1) \tag{16}$$

3. Variational Bayesian Approximation-Based Adaptive Kalman Filter Algorithm

As we know, when the measurement noise covariance is unknown or time varying, the classical approach of solving the problem is to use adaptive filters. The Bayesian approach is a typical choice. But it is usually hard to get the analytical form due to complex probability density function or high dimension of integration. In [27], variational Bayesian (VB) approximation was firstly used for the Kalman filter to estimate the joint posterior distribution of the states and noise covariances.

Given a discrete-time linear state space model as:

$$X_{k+1} = F_k X_k + w_k \tag{17}$$
$$y_k = H_k X_k + v_k \tag{18}$$

where $w_k \sim \mathcal{N}(0, Q_k)$ is a Gaussian distributed process noise, $v_k \sim \mathcal{N}(0, \Sigma_k)$ is a Gaussian measurement noise with diagonal covariance Σ_k. Note Q_k is assumed known but Σ_k is unknown.

Assuming the state vector and the measurement noise covariance are independent, the joint posterior distribution of the state and covariance $p(X_k, \Sigma_k | y_{1:k})$ can be solved by VB approximation as follows:

$$p(X_k, \Sigma_k | y_{1:k}) \approx Q_x(X_k) Q_\Sigma(\Sigma_k) \tag{19}$$

The VB approximation can now be formed by minimizing the Kullback–Leibler (KL) divergence between the separable approximation and the true posterior:

$$KL\left[Q_x(X_k) Q_\Sigma(\Sigma_k) || p(X_k, \Sigma_k | y_{1:k})\right]$$
$$= \int Q_x(X_k) Q_\Sigma(\Sigma_k) \times \log\left(\frac{Q_x(X_k) Q_\Sigma(\Sigma_k)}{p(X_k, \Sigma_k | y_{1:k})}\right) dX_k d\Sigma_k \tag{20}$$

Minimizing the KL divergence with respect to the probability densities $Q_x(X_k)$ and $Q_\Sigma(\Sigma_k)$ in turn, while keeping the other fixed, we can get the following equations:

$$Q_x(X_k) \propto exp\left(\int \log p(y_k, X_k, \Sigma_k | y_{1:k-1}) Q_\Sigma(\Sigma_k) d\Sigma_k\right) \tag{21}$$

$$Q_\Sigma(\Sigma_k) \propto exp\left(\int \log p(y_k, X_k, \Sigma_k | y_{1:k-1}) Q_x(X_k) dX_k\right) \tag{22}$$

Computing the above equations, we can get the following densities [27]:

$$Q_x(X_k) = \mathcal{N}(X_k | m_k, P_k) \tag{23}$$

$$Q_\Sigma(\Sigma_k) = \prod \text{Inv-Gamma}(\sigma_{k,i}^2 | \alpha_{k,i}, \beta_{k,i}) \tag{24}$$

where the parameters m_k, P_k, $\alpha_{k,i}$, $\beta_{k,i}$ can be calculated as:

$$
\begin{aligned}
m_k &= m_k^- + P_k^- H_k^T (H_k P_k^- H_k^T + \hat{\Sigma}_k)^{-1} (y_k - H_k m_k^-) \\
P_k &= P_k^- - P_k^- H_k^T (H_k P_k^- H_k^T + \hat{\Sigma}_k)^{-1} H_k P_k^- \\
\alpha_{k,i} &= 1/2 + \alpha_{k-1,i} \\
\beta_{k,i} &= \beta_{k-1,i} + \frac{1}{2}[(y_k - H_k m_k)_i^2 + (H_k P_k H_k^T)_{ii}]
\end{aligned}
\tag{25}
$$

where $i = 1, ..., d$ and d denote the dimensionality of the measurement. m_k^- and P_k^- are the predicted state estimate and its covariance, respectively. Here we assume each component of the measurement noise variance is mutually independent, and then the covariance matrix is diagonal, estimated as:

$$
\hat{\Sigma}_k = diag(\beta_{k,1}/\alpha_{k,1}, \ldots, \beta_{k,d}/\alpha_{k,d})
\tag{26}
$$

Furthermore, in order to describe the dynamics of the measurement noise variance, the inverse Gamma distribution parameters are assumed to change by a scale factor $\rho \in (0, 1]$. The formulas are given as:

$$
\begin{aligned}
\alpha_{k,i}^- &= \rho_i \alpha_{k-1,i} \\
\beta_{k,i}^- &= \rho_i \beta_{k-1,i}
\end{aligned}
\tag{27}
$$

Note that the value $\rho = 1$ corresponds to stationary variances and lower values represent larger time-fluctuations.

The above are the basic equations of adaptive Kalman filter based on VB approximation. Moreover, when the state equation and the measurement equation are nonlinear, the VB method can be rewritten in the EKF framework.

4. Variational Bayesian Approximation-Based Adaptive Dual Extended Kalman Filter

In order to handle the joint estimation of the SOC and the battery model parameters as well as the unknown measurement noise covariances, we propose a variational Bayesian approximation-based dual extended Kalman filter (VB-ADEKF) in this paper.

First, let us rewrite the state-space equations for SOC estimation as

$$
\left\{
\begin{aligned}
X_{k+1} &= F_k X_k + G_k I_{L,k} + w_k \\
y_k &= h(X_k, I_{L,k}, \theta_k) + v_k \\
w_k &\sim \mathcal{N}(0, Q_k^x) \\
v_k &\sim \mathcal{N}(0, \Sigma_k^x)
\end{aligned}
\right.
\tag{28}
$$

and the state-space equations for battery model parameters as

$$
\left\{
\begin{aligned}
\theta_{k+1} &= \theta_k + r_k \\
d_k &= h(X_k, I_{L,k}, \theta_k) + e_k \\
r_k &\sim N(0, Q_k^\theta) \\
e_k &\sim N(0, \Sigma_k^\theta)
\end{aligned}
\right.
\tag{29}
$$

where the process noise covariances Q_k^x, Q_k^θ are assumed known, and the measurement noise variances Σ_k^x, Σ_k^θ are unknown, being assumed as stochastic variables.

Then, VB-ADEKF is to alternatively solve the joint posterior distribution $p(X_k, \Sigma_k^x | y_{1:k})$ of the SOC and its measurement noise variance and the joint posterior distribution $p(\theta_k, \Sigma_k^\theta | y_{1:k})$ of the battery parameter and its measurement noise variance by VB approximation as follows:

$$p(X_k, \Sigma_k^x | y_{1:k}) \approx Q_x(X_k) Q_\Sigma(\Sigma_k^x) \tag{30}$$

$$p(\theta_k, \Sigma_k^\theta | y_{1:k}) \approx Q_\theta(\theta_k) Q_\Sigma(\Sigma_k^\theta) \tag{31}$$

where $Q_x(X_k)$, $Q_\Sigma(\Sigma_k^x)$, $Q_\theta(\theta_k)$, and $Q_\Sigma(\Sigma_k^\theta)$ can be solved by Equations (23) and (24). It should be noted that the parameters in these approximating densities are actually obtained using the extended Kalman filters since the measurement function $h(\cdot)$ is nonlinear. The filtering procedure of the VB-ADEKF algorithm is summarized in Algorithm 1.

Algorithm 1 : VB-ADEKF.

---

(1) Initialization: $\hat{X}_0$, $\hat{\theta}_0$, $P_{x,0}$, $P_{\theta,0}$, Q_0^x, Q_0^θ, $\hat{\alpha}_{x,0}$, $\hat{\beta}_{x,0}$, $\hat{\alpha}_{\theta,0}$, $\hat{\beta}_{\theta,0}$

(2) Prediction:

$$\hat{X}_k^- = F_{k-1}\hat{X}_{k-1} + G_{k-1}I_{L,k-1}, \quad \hat{\theta}_k^- = \hat{\theta}_{k-1}$$

$$P_{x,k}^- = F_{k-1}P_{x,k-1}F_{k-1}^T + Q_k^x, \quad P_{\theta,k}^- = P_{\theta,k-1} + Q_k^\theta$$

$$\hat{\alpha}_{x,k}^- = \rho_x\hat{\alpha}_{x,k-1}, \quad \hat{\beta}_{x,k}^- = \rho_x\hat{\beta}_{x,k-1}$$

$$\hat{\alpha}_{\theta,k}^- = \rho_\theta\hat{\alpha}_{\theta,k-1}, \quad \hat{\beta}_{\theta,k}^- = \rho_\theta\hat{\beta}_{\theta,k-1}$$

where α_x, β_x and α_θ, β_θ are the inverse gamma distribution parameters of the measurement noise covariance, ρ_x and ρ_θ are the scale factors.

(3) Update: the update of VB-ADEKF utilizes iterate filtering framework.

First set: $\hat{X}_k^{(0)} = \hat{X}_k^-$, $P_{x,k}^{(0)} = P_{x,k}^-$, $\theta_k^{(0)} = \hat{\theta}_k^-$, $P_{\theta,k}^{(0)} = P_{\theta,k}^-$

$$\hat{\alpha}_{x,k} = 1/2 + \hat{\alpha}_{x,k}^-, \quad \hat{\beta}_{x,k}^{(0)} = \hat{\beta}_{x,k}^- \quad \hat{\alpha}_{\theta,k} = 1/2 + \hat{\alpha}_{\theta,k}^-, \quad \hat{\beta}_{\theta,k}^{(0)} = \hat{\beta}_{\theta,k}^-$$

For $n = 1 : N$, iterate the following N (N denotes iterated times) steps:

- **Measurement variances:**
$$\hat{\Sigma}_{x,k}^{(n)} = \hat{\beta}_{x,k}^{(n)}/\hat{\alpha}_{x,k}^{(n)}, \quad \hat{\Sigma}_{\theta,k}^{(n)} = \hat{\beta}_{\theta,k}^{(n)}/\hat{\alpha}_{\theta,k}^{(n)}$$

- **State estimate and its covariance:**
$$\hat{X}_k^{(n+1)} = \hat{X}_k^- + P_{x,k}^- H_{x,k}^T (H_{x,k}P_{x,k}^- H_{x,k}^T + \hat{\Sigma}_{x,k}^{(n)})^{-1}(y_k - h(\hat{X}_k^-, I_{L,k}, \hat{\theta}_k^-))$$

$$P_{x,k}^{(n+1)} = P_{x,k}^- - P_{x,k}^- H_{x,k}^T (H_{x,k}P_{x,k}^- H_{x,k}^T + \hat{\Sigma}_{x,k}^{(n)})^{-1} H_{x,k}P_{x,k}^-$$

- **Battery parameters estimate and its covariance:**
$$\hat{\theta}_k^{(n+1)} = \hat{\theta}_k^- + P_{\theta,k}^- H_{\theta,k}^T (H_{\theta,k}P_{\theta,k}^- H_{\theta,k}^T + \hat{\Sigma}_{\theta,k}^{(n)})^{-1}(y_k - h(\hat{X}_k^-, I_{L,k}, \hat{\theta}_k^-))$$

$$P_{\theta,k}^{(n+1)} = P_{\theta,k}^- - P_{\theta,k}^- H_{\theta,k}^T (H_{\theta,k}P_{\theta,k}^- H_{\theta,k}^T + \hat{\Sigma}_{\theta,k}^{(n)})^{-1} H_{\theta,k}P_{\theta,k}^-$$

- **Parameters for the measurement noise variances estimation:**
$$\hat{\beta}_{x,k}^{(n+1)} = \hat{\beta}_{x,k}^- + \tfrac{1}{2}\left(y_k - h(\hat{X}_k^{(n+1)}, I_{L,k}, \hat{\theta}_k^{(n+1)})\right)^2 + \tfrac{1}{2}H_{x,k}P_{x,k}^{(n+1)}H_{x,k}^T$$

$$\hat{\beta}_{\theta,k}^{(n+1)} = \hat{\beta}_{\theta,k}^- + \tfrac{1}{2}\left(y_k - h(\hat{X}_k^{(n+1)}, I_{L,k}, \hat{\theta}_k^{(n+1)})\right)^2 + \tfrac{1}{2}H_{\theta,k}P_{\theta,k}^{(n+1)}H_{\theta,k}^T$$

End for.

And set $\hat{\beta}_{x,k} = \hat{\beta}_{x,k}^{(N)}$, $\hat{\beta}_{\theta,k} = \hat{\beta}_{\theta,k}^{(N)}$, $\hat{X}_k = \hat{X}_k^{(N)}$, $P_{x,k} = P_{x,k}^{(N)}$, $\hat{\theta}_k = \hat{\theta}_k^{(N)}$, $P_{\theta,k} = P_{\theta,k}^{(N)}$

By alternatively using two VB-based extended Kalman filters for online estimation of the battery SOC and model parameters, while compensating for the uncertainties in the model parameters by simultaneous estimation of the measurement noise variances, the adaptability of the proposed algorithm to dynamic changes in battery characteristics is greatly improved. Hence, it is very promising to further increase the SOC estimation accuracy and robustness.

5. Experimental Verification and Analysis

5.1. Experimental Settings

The proposed method is experimentally evaluated in this section. The experimental setup for the tests is shown in Figure 2. The setup consists of three lithium-ion battery cells connected in series, an electronic load, and a host computer. The tested lithium-ion battery cells are type 18,650, whose nominal capacity is 2200 mAh, nominal voltage is 3.7 V, charging and discharging cutoff voltages are 4.2 V and 3 V, respectively. The type of the electronic load is IT8516S produced by ITECH, whose current measurement accuracy is $\pm(0.1\% + 0.1\%$ full scale), and voltage measurement accuracy is $\pm(0.02\% + 0.02\%$ full scale). The load current, terminal voltage, and SOC can be recorded via the host computer during the discharge test.

First, a sequence of pulsed discharging experiments were implemented to determine the relationship of open circuit voltage (OCV) and SOC. By using the curve-fitting toolbox in MATLAB, k_0, k_1, k_2, k_3, k_4, k_5 were identified, which are shown in Table 1. The measured data and fitted curve are presented in Figure 3. The R-square is used to represent the goodness of fit. The normal value range of the R-square is 0–1 and a value closer to 1 indicates a better fitting curve [32]. It can be seen that the curve fits well with the measurement data, indicating that the selected fifth-order polynomial model can describe the relationship of OCV and SOC very well.

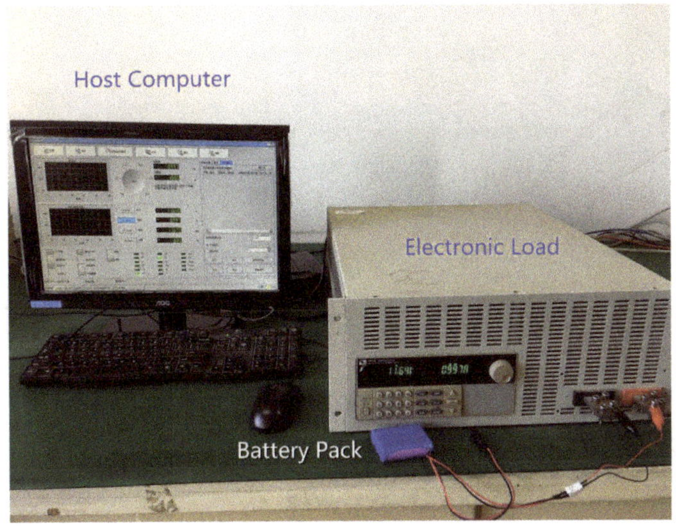

Figure 2. The experimental setup.

Table 1. The identification results of k_0, k_1, k_2, k_3, k_4, k_5.

k_0	k_1	k_2	k_3	k_4	k_5
9.50	14.01	−62.64	137.40	−133.40	47.76

Then, a constant current discharge test, a pulse current discharge test, and an urban dynamometer driving schedule (UDDS) test were performed. They are commonly used to verify the performances of SOC estimation methods in EVs. In the constant current discharge test, the current keeps invariant but the terminal voltage declines continuously. In the pulse current discharge test, the current stays at 1 A for 10 min and then decreases to 0 and lasts for 30 min. The process is repeated until the battery reaches the lower cut-off voltage. UDDS, also known as FTP72, was used as a test procedure to certify vehicle emissions by the US Environmental Protection Agency. It can simulate the actual driving conditions of vehicles on the road. The battery current and voltage are both sampled at 1 s. In each test, the true SOC was obtained using CC method. The estimation accuracy, convergence rate, and robustness of the proposed VB-ADEKF are evaluated by comparison with DEKF under different tests.

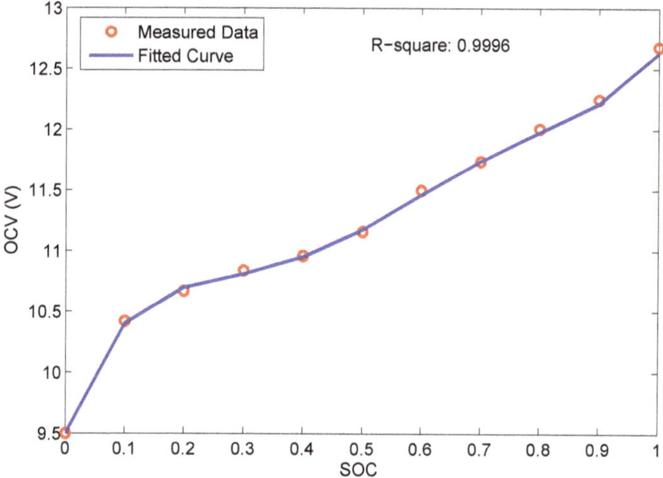

Figure 3. The relationship curve of open circuit voltage (OCV) versus state of charge (SOC).

5.2. Constant Current Discharge Test

The experiment was performed with a constant discharge current of 1 A. The initial SOC value is set to be 0.8, rather than the real SOC of 1.0. The process noise covariances are set as $Q_k^x = 1 \times 10^{-6} I_2$ and $Q_k^\theta = 1 \times 10^{-6} I_3$. The measurement noise variances used for DEKF are $\Sigma_k^x = 0.001$ and $\Sigma_k^\theta = 0.0005$. The scale factors for VB-ADEKF are set to 1×10^{-4}. The initial parameters α_0 and β_0 for battery parameters and SOC are both set as 10 and 0.001, respectively.

The estimated values of the battery model parameters by VB-ADEKF are presented in Figure 4. The ohmic resistance is stable at the beginning of the discharge and increases at the end of the discharge. The polarization resistance decreases at first and increases later with the depth of the discharge and again decreases in the last 1000 s. The polarization capacitance shows a declining trend overall in the process of discharge.

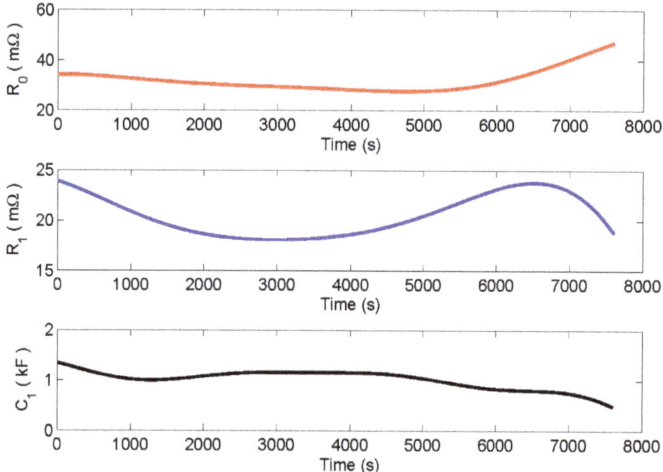

Figure 4. Results of parameter identification using variational Bayesian approximation-based adaptive dual extended Kalman filter (VB-ADEKF) in the constant current discharge test.

Furthermore, to verify the accuracy of online identification of the battery parameters by the VB-ADEKF algorithm, the measured terminal voltage and the estimated terminal voltage by VB-ADEKF were compared, as shown in Figure 5. The maximum absolute error was 0.023 V, except for the first big error caused by an incorrect initial SOC value. The mean absolute error was 0.0050 V and the relative mean absolute error was 0.052%. It is clear that the estimated terminal voltage agrees well with the measured voltage. This illustrates the effectiveness of the battery model whose parameters are identified by the proposed VB-ADEKF method.

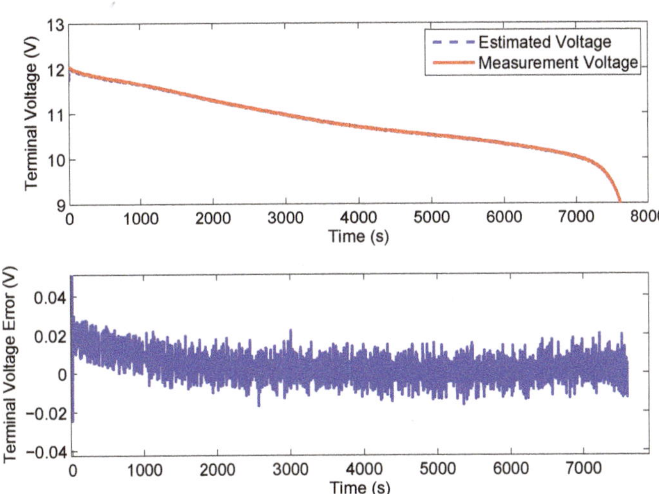

Figure 5. Experimental terminal voltage results using VB-ADEKF in the constant current discharge test.

The SOC estimation results using VB-ADEKF and DEKF are shown in Figure 6. Clearly, VB-ADEKF has a much more accurate SOC estimation than DEKF. The SOC estimation error of VB-ADEKF is bounded within ±1% for most of the time, but DEKF goes outside of this interval. The detailed error values are shown in Table 2. The maximum absolute estimation error of VB-ADEKF is 1.28% and the mean absolute error is 0.64% after discharge for 12 min. Both are significantly smaller than the errors of DEKF, which are correspondingly 2.76% and 1.39%. Meanwhile, from the figure it can be seen that VB-ADEKF converges much faster than DEKF under initial SOC errors. This is also verified in Table 2. The convergence time, which is defined as the first time instant at which the SOC estimation error decreases to ±5%, are 10 s and 335 s for VB-ADEKF and DEKF, respectively.

Table 2. Statistical analysis of state of charge (SOC) estimation error (after 12 min) in three tests.

	Constant Current Test		Pulse Current Test		UDDS Test	
	DEKF	**VB-ADEKF**	**DEKF**	**VB-ADEKF**	**DEKF**	**VB-ADEKF**
Maximum Absolute Error	2.76%	1.28%	4.93%	5.00%	4.72%	4.10%
Mean Absolute Error	1.39%	0.64%	1.01%	0.68%	1.26%	0.89%
Convergence Time	335 s	10 s	698 s	690 s	675 s	603 s

5.3. Pulse Current Discharge Test

In the pulse current discharge test, the initial SOC value is set to be 0.8 while the real SOC is 1.0. The process noise covariances are set as $Q_k^x = 1 \times 10^{-6} I_2$ and $Q_k^\theta = 1 \times 10^{-6} I_3$. The measurement noise variances used for DEKF are $\Sigma_k^x = 0.01$ and $\Sigma_k^\theta = 0.0005$. The initial parameter values of VB-ADEKF are set as $\alpha_{\theta,0} = 100$, $\beta_{\theta,0} = 0.0005$, $\alpha_{x,0} = 10$, $\beta_{x,0} = 0.001$.

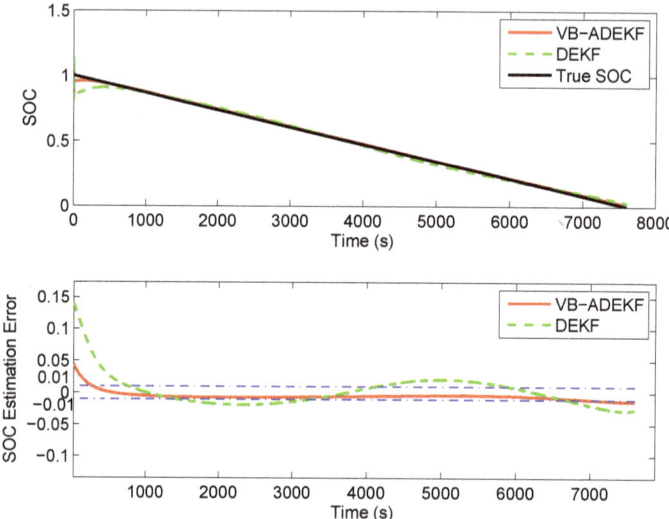

Figure 6. SOC estimation results using VB-ADEKF and dual extended Kalman filter (DEKF) in the constant current discharge test.

Figure 7 presents the estimated values of the battery model parameters. Clearly, there is a stepped increase in the ohmic resistance from 34 mΩ to 49 mΩ. The polarization resistance retains stable during the entire discharge process. The polarization capacitance first decreases rapidly then stays stable until the end of the discharge. Figure 8 shows the the measured terminal voltage and the estimated terminal voltage by VB-ADEKF to verify the effectiveness of the battery parameters identification. The maximum and mean absolute estimation errors are 0.086 V and 0.0027 V, respectively. It implies that the estimated terminal voltage has a good agreement with the measured voltage. It further shows the battery model parameters are well identified.

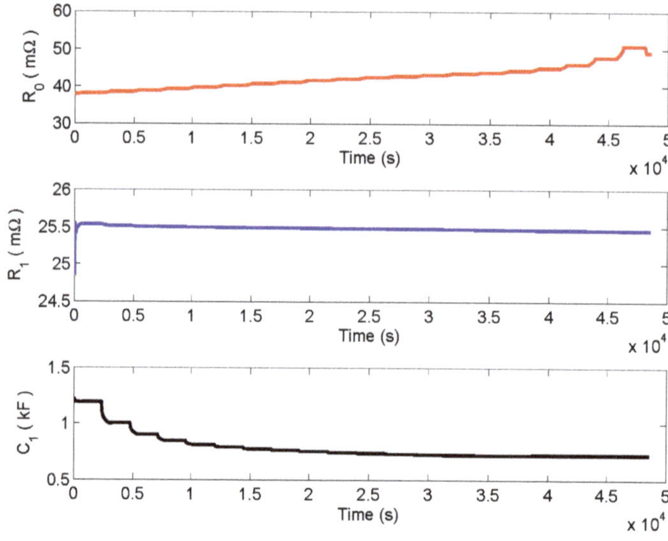

Figure 7. Results of parameters identification using VB-ADEKF in the pulse current discharge test.

The SOC estimation results of VB-ADEKF and DEKF are plotted in Figure 9. It shows that VB-ADEKF and DEKF have comparable performance in convergence rate. However VB-ADEKF has more accurate SOC estimation than that of DEKF. This is also verified in Table 2, in which the mean absolute error of VB-ADEKF is 0.68%, while it is 1.01% for DEKF after discharge for 12 min.

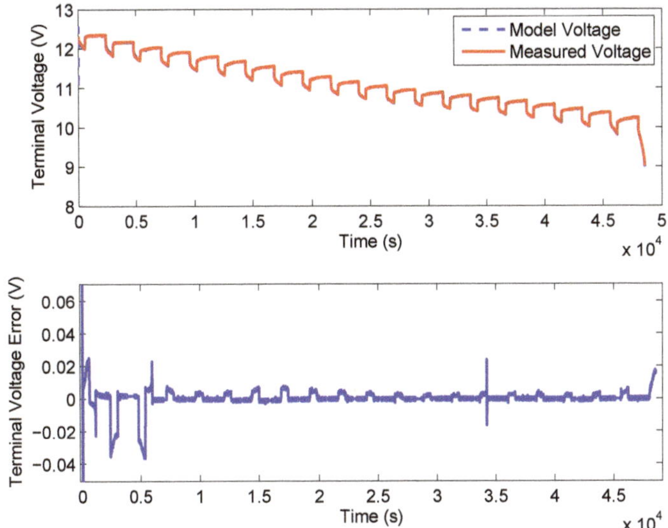

Figure 8. Experimental terminal voltage results using VB-ADEKF in the pulse current discharge test.

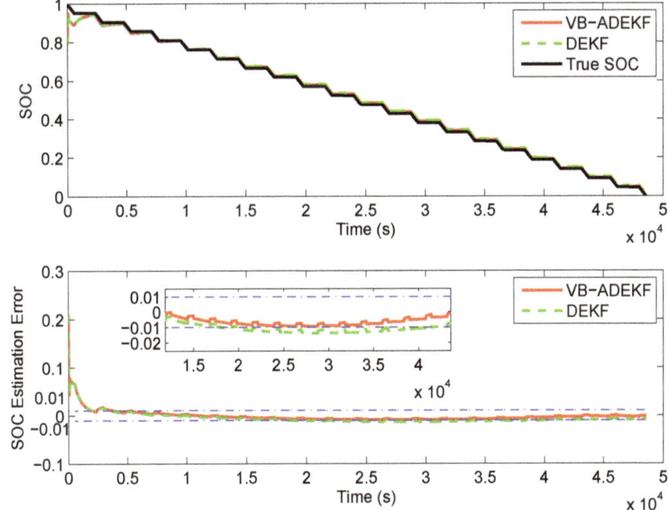

Figure 9. SOC estimation results using VB-ADEKF and DEKF in the pulse current discharge test.

5.4. UDDS Test

To evaluate the SOC estimation performance under dynamic loading profiles, an UDDS cycle was performed on the battery cells. According to the actual tolerable currents of the lithium-ion battery cells, the loading currents are scaled down, as shown in Figure 10. The initial SOC value is set to 0.8.

The process noise covariances are set as $Q_k^x = 1 \times 10^{-6} I_2$ and $Q_k^\theta = 1 \times 10^{-6} I_3$. The measurement noise variances used for DEKF are $\Sigma_k^x = 0.01$ and $\Sigma_k^\theta = 0.0005$. The initial parameter values of VB-ADEKF are set as $\alpha_{\theta,0} = 100$, $\beta_{\theta,0} = 0.0005$, $\alpha_{x,0} = 10$, $\beta_{x,0} = 0.001$.

The estimated values of the battery model parameters are presented in Figure 11. It can be seen that the range of parameter values of R_0, R_1 and C_1 are consistent with these in constant current discharge test and pulse current discharge test, but the changing rules are slightly different. Figure 12 presents the the measured terminal voltage and the estimated terminal voltage by VB-ADEKF to verify the effectiveness of the battery parameters identification. The maximum and mean absolute estimation errors are 0.062 V and 0.0011 V, respectively. It is clear that the estimated terminal voltage has a good consistency with the measured voltage. It further shows the battery model parameters are well identified.

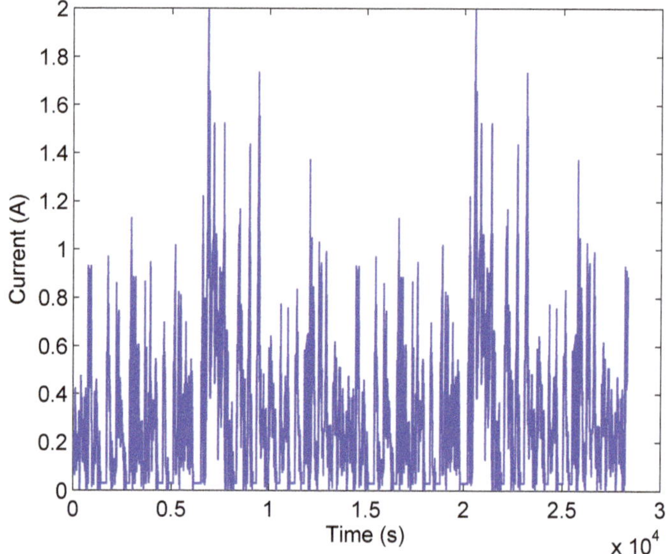

Figure 10. Current profiles in the urban dynamometer driving schedule (UDDS) test.

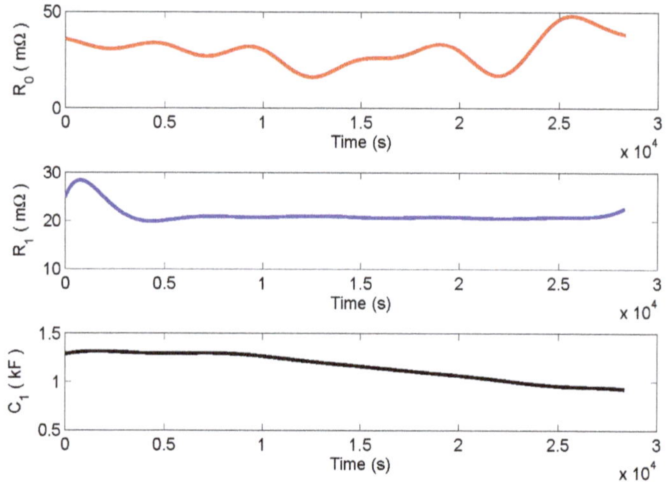

Figure 11. Results of parameters identification using VB-ADEKF in the UDDS test.

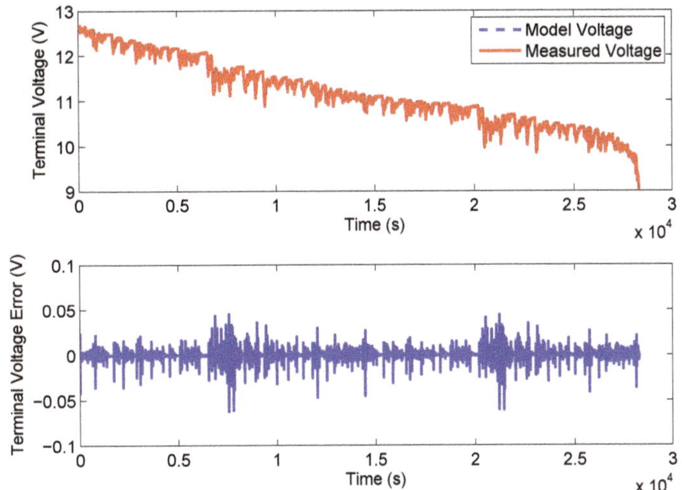

Figure 12. Experimental terminal voltage results using VB-ADEKF in the UDDS test.

Figure 13 presents the SOC estimation results of VB-ADEKF and DEKF. It shows that both VB-ADEKF and DEKF have good estimation accuracy when SOC is between 20% and 90%. But VB-ADEKF still outperforms the traditional DEKF in SOC estimation accuracy and convergence rate. A comparative summary of the two methods is given in Table 2. When SOC is decreased to 20%, the estimation error of both methods begins to increase, but is no larger than 5%. This may be because the polarization effect of the battery is further aggravated at lower SOC level. As a result, the terminal voltage measurement error has increased, thereby reducing the SOC estimation accuracy.

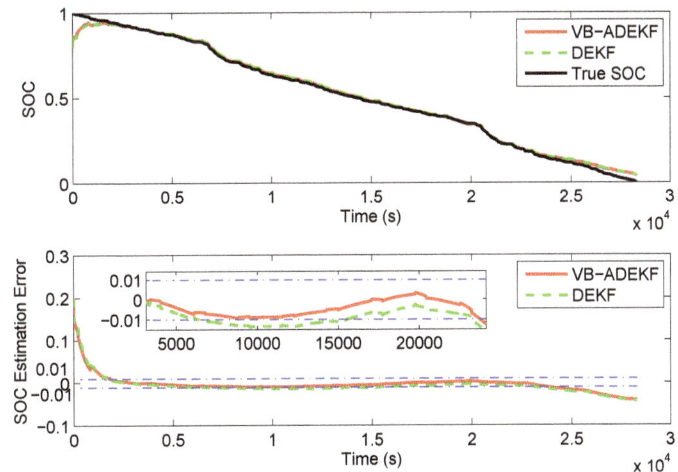

Figure 13. SOC estimation results using VB-ADEKF and DEKF in the UDDS test.

5.5. Convergence Ability with Initial SOC Error

Because it is difficult to determine the initial SOC value precisely in practical applications, it is important and indispensable for the algorithms to have the ability to correct the uncertainty brought by the initial SOC error. Therefore, the convergence rate with initial SOC error is adopted as another indicator for evaluating the SOC estimation algorithms. The true initial SOC value is 1.0. Figures 14–16 present the estimation results of VB-ADEKF and DEKF algorithms for different initial SOC values from 30% to 90% under the above three tests, respectively. Overall, the convergence time increases as the initial SOC error rises for both VB-ADEKF and DEKF. But the growth rates are very different for the two filters under different tests.

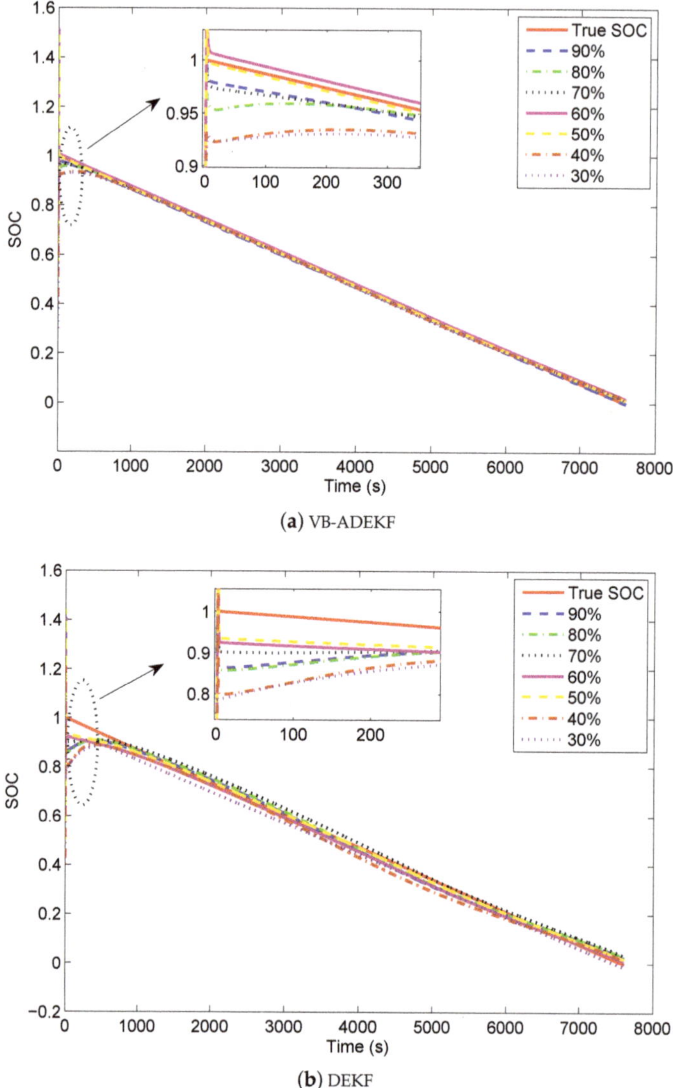

Figure 14. SOC estimation results of the two filters with different initial SOC values in the constant current discharge test.

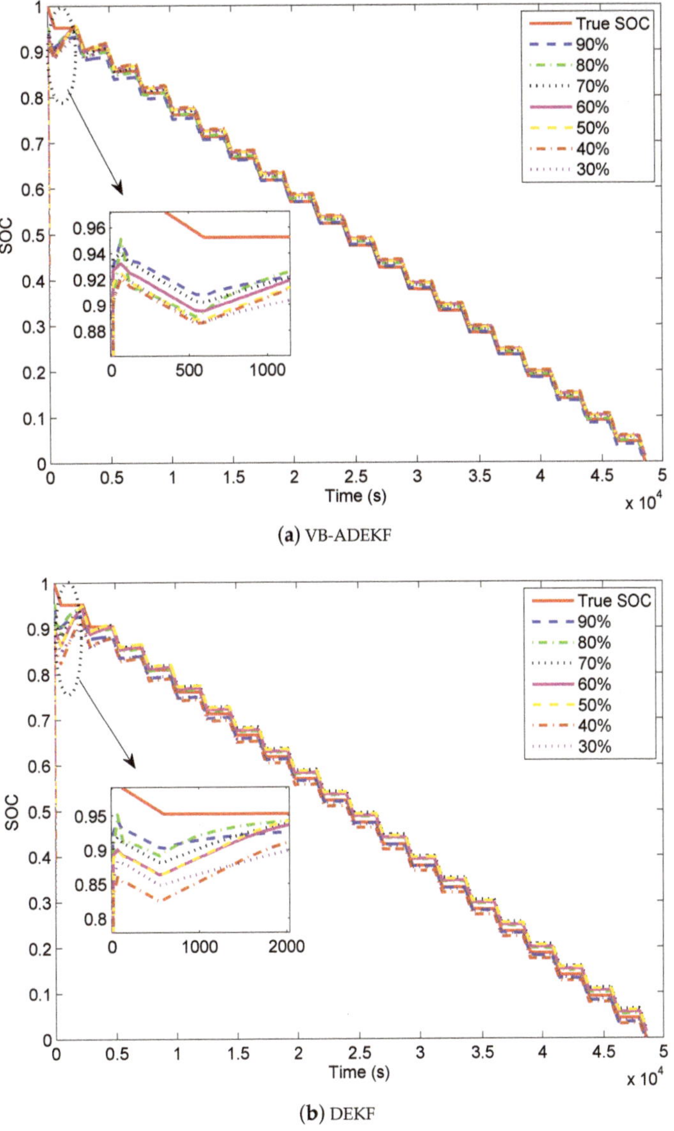

(a) VB-ADEKF

(b) DEKF

Figure 15. SOC estimation results of the two filters with different initial SOC values in the pulse current discharge test.

Specifically, from the results of the convergence time shown in Table 3, it can be seen that the convergence time of VB-ADEKF is stabilized about 10 s into the constant current discharge test when the initial SOC value is larger than 50%, and increased to more than 100 s when the initial SOC values are 40% and 30%. But the convergence time of DEKF, which is between 300 s and 500 s, is much longer than VB-ADEKF. It shows that VB-ADEKF can quickly converge to the true SOC values without resulting in the accumulation of errors caused by the initial error of the SOC. In the pulse current discharge test, the convergence rates are comparable for VB-ADEKF and DEKF in the case of small initial SOC errors, for example, 10% or 20%. However, when the initial SOC error is relatively large, the convergence time of VB-ADEKF becomes much smaller than that of DEKF. In the UDDS test, VB-ADEKF only exhibits a slightly increasing trend with the increase of the initial SOC error. However, DEKF converges slower than VB-ADEKF, in the meantime, with a larger SOC estimation error. Its convergence time goes up quickly as the initial SOC error increases. This implies that the initial SOC error has a noticeable impact on the performance of DEKF. But, from the overall perspective, VB-ADEKF is not very sensitive to the initial SOC error. It shows that the proposed VB-ADEKF has better robustness for initial SOC errors than the traditional DEKF.

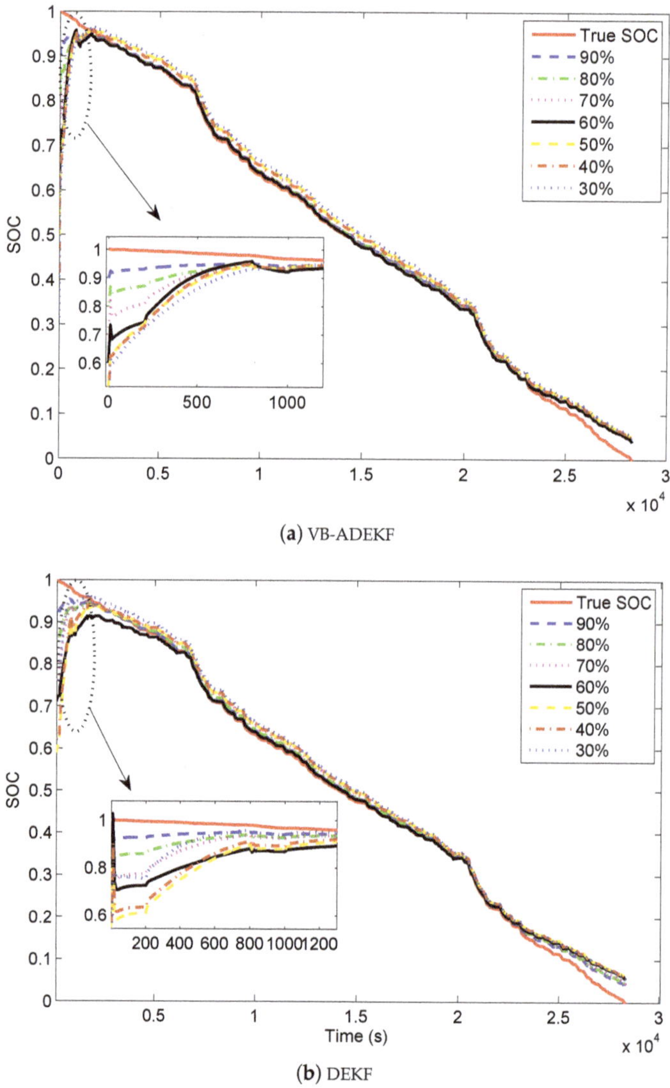

(**a**) VB-ADEKF

(**b**) DEKF

Figure 16. SOC estimation results of the two filters with different initial SOC values in the UDDS test.

Table 3. Convergence time (s) of variational Bayesian approximation-based adaptive dual extended Kalman filter (VB-ADEKF) and dual extended Kalman filter (DEKF) with different initial SOC values.

		Initial SOC Values						
		90%	80%	70%	60%	50%	40%	30%
Constant current test	DEKF	341	335	331	425	262	468	483
	VB-ADEKF	2	10	9	10	12	135	157
Pulse current test	DEKF	596	698	1020	1253	1238	1820	2121
	VB-ADEKF	410	690	629	776	918	943	1115
UDDS test	DEKF	383	675	718	1133	1420	1890	2085
	VB-ADEKF	359	603	610	612	695	658	783

5.6. Effect of Mistuning

If the working condition of the battery changes abruptly, the SOC measurement error would probably be varied largely with before, so the prior tuning of the measurement variance of the DEKF will not give an optimal estimate of the SOC. But there is no such issue in the proposed VB-ADEKF since the measurement variances are estimated online. This effects of mistuning brought about by inappropriate measurement variance of the traditional DEKF and VB-ADEKF are mainly reflected in the SOC estimation error, as shown in Figures 17–19. Here, the measurement noise variances of DEKF are mistuned to $\Sigma_k^x = 0.01$ and $\Sigma_k^\theta = 0.001$ in constant current discharge test, and $\Sigma_k^x = 0.1$ and $\Sigma_k^\theta = 0.005$ in the pulse current discharge test and UDDS test. The initial estimates of the measurement variances of VB-ADEKF are also correspondingly mistuned. From the results, we can see that the convergence rate of DEKF slows down and the SOC estimation accuracy also declines in the case of mistuning, while the SOC estimation performance of the proposed VB-ADEKF remains almost unchanged. This suggests that the proposed VB-ADEKF is more robust than the traditional DEKF.

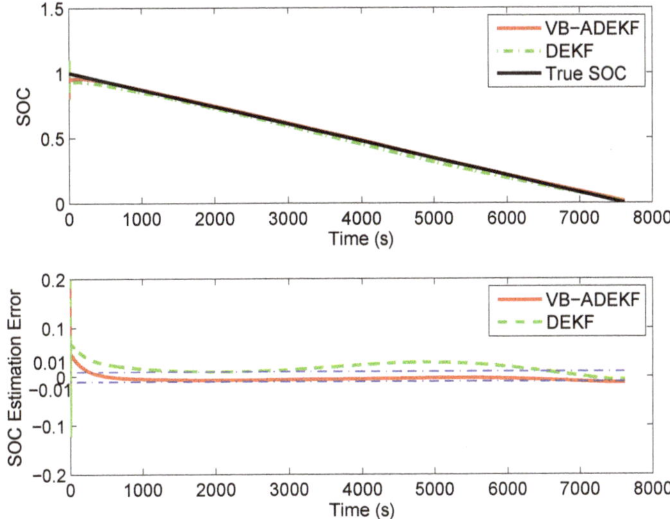

Figure 17. SOC estimation results using VB-ADEKF and DEKF in the case of mistuning in the constant current discharge test.

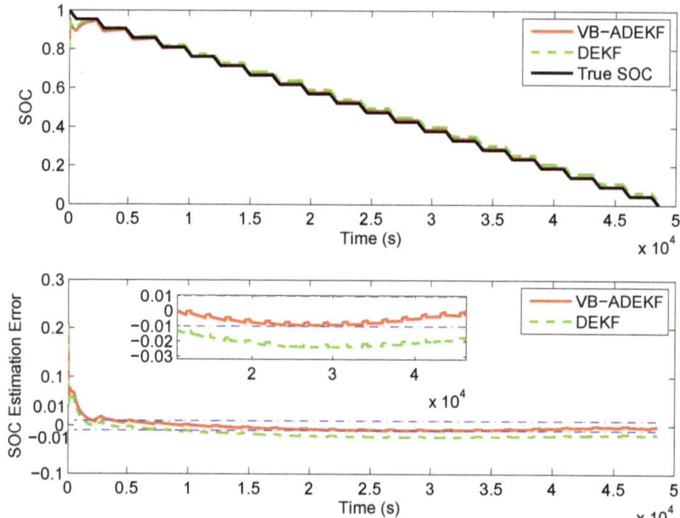

Figure 18. SOC estimation results using VB-ADEKF and DEKF in the case of mistuning in the pulse current discharge test.

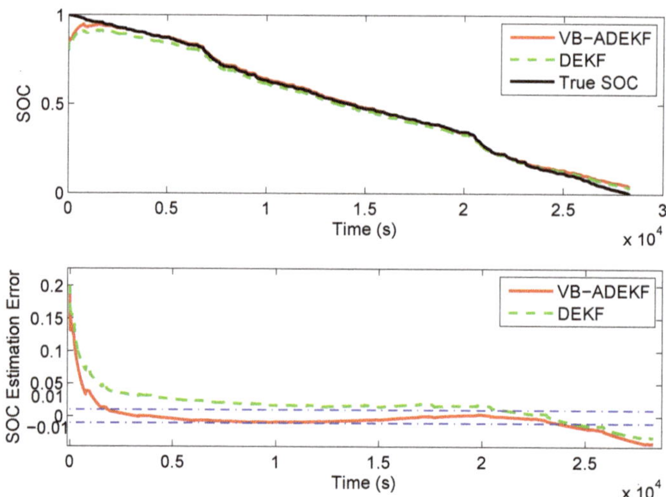

Figure 19. SOC estimation results using VB-ADEKF and DEKF in the case of mistuning in the UDDS test.

6. Conclusions

To deal with the measurement statistical uncertainties and inaccurate battery model, a variational Bayesian approximation-based adaptive dual extended Kalman filter (VB-ADEKF) is proposed in this paper for SOC estimation of lithium-ion batteries. First, the variational Bayesian inference is integrated with the extended Kalman filter to jointly estimate the states and the measurement noise covariances. Then, two VB-based extended Kalman filters are alternatively used for online estimation of the battery SOC and model parameters, while simultaneously estimating the measurement noise variances to compensate for the uncertainties in the measurement and battery parameters. Therefore, the adaptability of the proposed algorithm to dynamic changes in battery characteristics is greatly improved. The effectiveness and superiority of the proposed algorithm have been verified by comparing with the dual EKF (DEKF) algorithm through experiments under the constant current discharge test, pulse current discharge test, and UDDS test. The results show that the proposed VB-ADEKF algorithm outperforms the traditional DEKF approach in terms of SOC estimation accuracy and convergence rate. Especially, when the quality of measurements changes with the operating conditions, the proposed VB-ADEKF exhibits better robustness than DEKF.

Author Contributions: J.H. conceived this paper, designed the experiments, and analyzed the data; T.G. and H.H. performed the experiments; Y.Y. revised the paper and provided some valuable suggestions.

Funding: This research was funded by Shaanxi Provincial Key Research and Development Programs (No. 2017ZDXM-GY-06, No. 2017GY-057), and Xi'an Science and Technology Planning Project–Scientific and Technological Innovation Guidance Project (No. 201805042YD20CG26 (8)).

Conflicts of Interest: The authors declare no conflict of interest.

Abbreviations

The following abbreviations are used in this manuscript:

AEKF	Adaptive Extended Kalman Filter
ANFIS	Adaptive Neuro-Fuzzy Inference System
ALS	Autocovariance Least Squares
ASRUKF	Adaptive Square Root Unscented Kalman Filter
AUKF	Adaptive Unscented Kalman Filter
BMS	Battery Management System
CC	Coulomb Counting
CKF	Cubature Kalman Filter
DEKF	Dual Extended Kalman Filter
EKF	Extended Kalman Filter
EV	Electric Vehicle
FFRLS	Forgetting-Factor Recursive Least-Squares
FL	Fuzzy Logic
KF	Kalman Filter
KL	Kullback–Leibler
MM	Multiple Model
NN	Neural Network
OCV	Open Circuit Voltage
RC	Resistor–Capacitor
SOC	State Of Charge
SVM	Support Vector Machine
UDDS	Urban Dynamometer Driving Schedule
UKF	Unscented Kalman Filter
VB	Variational Bayesian

References

1. Pan, H.; Lü, Z.; Lin, W.; Li, J.; Chen, L. State of charge estimation of lithium-ion batteries using a grey extended Kalman filter and a novel open-circuit voltage model. *Energy* **2017**, *138*, 764–775. [CrossRef]
2. Hannan, M.A.; Lipu, M.S.H.; Hussain, A.; Mohamed, A. A review of lithium-ion battery state of charge estimation and management system in electric vehicle applications: Challenges and recommendations. *Renew. Sustain. Energy Rev.* **2017**, *78*, 834–854. [CrossRef]
3. He, W.; Williard, N.; Chen, C.; Pecht, M. State of charge estimation for Li-ion batteries using neural network modeling and unscented Kalman filter-based error cancellation. *Int. J. Electr. Power Energy Syst.* **2014**, *62*, 783–791. [CrossRef]

4.   Zhao, W.; Kong, X.; Wang, C. Combined estimation of the state of charge of a lithium battery based on a back-propagation- adaptive Kalman filter algorithm. *Proc. Inst. Mech. Eng. Part D J. Automob. Eng.* **2018**, *232*, 357–366. [CrossRef]

5.   Charkhgard, M.; Farrokhi, M. State-of-charge estimation for lithium-ion batteries using neural networks and EKF. *IEEE Trans. Ind. Electron.* **2010**, *57*, 4178–4187. [CrossRef]

6.   Singh, P.; Vinjamuri, R.; Wang, X.; Reisner, D. Design and implementation of a fuzzy logic-based state-of-charge meter for Li-ion batteries used in portable defibrillators. *J. Power Sources* **2006**, *162*, 829–836. [CrossRef]

7.   Cai, C.H.; Du, D.; Liu, Z.Y. Battery state-of-charge (SOC) estimation using adaptive neuro-fuzzy inference system (ANFIS). In Proceedings of the 12th IEEE International Conference on Fuzzy Systems, St Louis, MO, USA, 25–28 May **2003**; pp. 1068–1073.

8.   Awadallah, M.A.; Venkatesh, B. Accuracy improvement of SOC estimation in lithium-ion batteries. *J. Energy Storage* **2016**, *6*, 95–104. [CrossRef]

9.   Hu, J.N.; Hu, J.J.; Lin, H.B.; Li, X.P.; Jiang, C.L.; Qiu, X.H.; Li, W.S. State-of-charge estimation for batterymanagement system using optimized support vectormachine for regression. *J. Power Sources* **2014**, *269*, 682–693. [CrossRef]

10.   Sheng, H.; Xiao, J. Electric vehicle state of charge estimation: Nonlinear correlation and fuzzy support vector machine. *J. Power Sources* **2015**, *281*, 131–137. [CrossRef]

11.   Wu, X.; Mi, L.; Tan, W.; Qin, J.L.; Zhao, M.N. State of charge (SOC) estimation of Ni-MH battery based on least square support vector machines. *Adv. Mater. Res.* **2011**, *211*, 1204–1209. [CrossRef]

12.   Chen, Z.; Fu, Y.; Mi, C.C. State of charge estimation of lithium-ion batteries in electric drive vehicles using extended Kalman filtering. *IEEE Trans. Veh. Technol.* **2013**, *62*, 1020–1030. [CrossRef]

13.   He, H.; Xiong, R.; Zhang, X.; Sun, F.; Fan, J. State-of-charge estimation of the Lithium-ion battery using an adaptive extended Kalman filter based on an improved Thevenin model. *IEEE Trans. Veh. Technol.* **2011**, *60*, 1461–1469.

14.   He, W.; Williard, N.; Chen, C.; Pecht, M. State of charge estimation for electric vehicle batteries using unscented kalman filtering. *Microelectron. Reliab.* **2013**, *53*, 840–847. [CrossRef]

15.   Li, Y.; Wang, C.; Gong, J. A multi-model probability SOC fusion estimation approach using an improved adaptive unscented Kalman filter technique. *Energy* **2017**, *141*, 1402–1415. [CrossRef]

16.   Peng, S.; Chen, C.; Shi, H.; Yao, Z. State of charge estimation of battery energy storage systems based on adaptive unscented Kalman filter with a noise statistics estimator. *IEEE Access* **2017**, *5*, 13202–13212. [CrossRef]

17.   Zeng, Z.; Tian, J.; Li, D.; Tian, Y. An online state of charge estimation algorithm for lithium-ion batteries using an improved adaptive cubature Kalman filter. *Energies* **2018**, *11*, 59. [CrossRef]

18.   Xiong, R.; Zhang, Y.; He, H.; Zhou, X.; Pecht, M. A double-scale, particle filtering, energy state prediction algorithm for lithium-ion batteries. *IEEE Trans. Ind. Electron.* **2018**, *65*, 1526–1538. [CrossRef]

19.   Wang, Y.; Zhang, C.; Chen, Z. A method for state-of-charge estimation of LiFePO4 batteries at dynamic currents and temperatures using particle filter. *J. Power Sources* **2015**, *279*, 306–311. [CrossRef]

20.   Nejad, S.; Gladwin, D.T.; Stone, D.A. On-chip implementation of extended kalman filter for adaptive battery states monitoring. In Proceedings of the 42nd Annual Conference of the IEEE Industrial Electronics Society, Florence, Italy, 23–26 October 2016; pp. 5513–5518.

21.   Cai, M.; Chen, W.; Tan, X. Battery state-of-charge estimation based on a dual unscented Kalman filter and fractional variable-order model. *Energies* **2017**, *10*, 1577. [CrossRef]

22.   Zhang, Y.; Zhang, C.; Zhang, X. State-of-charge estimation of the lithium-ion battery system with time-varying parameter for hybrid electric vehicles. *Control Theory Appl. IET* **2014**, *8*, 160–167. [CrossRef]

23.   Charkhgard, M.; Zarif, M.H. Design of adaptive H∞ filter for implementing on state of- charge estimation based on battery state-of-charge-varying modelling. *IET Power Electron.* **2015**, *8*, 1825–1833. [CrossRef]

24.   Yu, Q.; Xiong, R.; Lin, C.; Shen, W.; Deng, J. Lithium-ion battery parameters and state-of-charge joint estimation based on H-infinity and unscented Kalman filters. *IEEE Trans. Veh. Technol.* **2017**, *66*, 8693–8701. [CrossRef]

25.   Liu, S.; Cui, N.; Zhang, C. An adaptive square root unscented Kalman filter approach for state of charge estimation of lithium-ion batteries. *Energies* **2017**, *10*, 1345.

26. El Din, M.S.; Abdel-Hafez, M.F.; Hussein, A.A. Enhancement in Li-ion battery cell state-of-charge estimation under uncertain model statistics. *IEEE Trans. Veh. Technol.* **2016**, *65*, 4608–4618. [CrossRef]

27. Sarkka, S.; Nummenmaa, A. Recursive noise adaptive Kalman filtering by variational Bayesian approximations. *IEEE Trans. Autom. Control* **2009**, *54*, 596–600. [CrossRef]

28. Li, K.; Chang, L.; Hu, B. A variational Bayesian-based unscented Kalman filter with both adaptivity and robustness. *IEEE Sens. J.* **2016**, *16*, 6966–6976. [CrossRef]

29. Sun, J.; Zhou, J.; Li, X.R. State estimation for systems with unknown inputs based on variational Bayes method. In Proceedings of the 15th International Conference on Information Fusion, Singapore, 9–12 July 2012; pp. 983–990.

30. Hou, J.; Yang, Y.; Gao, T. Variational Bayesian based adaptive shifted Rayleigh filter for bearings-only tracking in clutters. *Sensors* **2019**, *19*, 1512. [CrossRef] [PubMed]

31. Zhao, Y.; Fatehi, A.; Huang, B. Robust estimation of ARX models with time varying time delays using variational Bayesian approach. *IEEE Trans. Cybern.* **2018**, *48*, 532–542. [CrossRef]

32. Wang, Q.; Kang, J.; Tan, Z.; Luo, M. An online method to simultaneously identify the parameters and estimate states for lithium ion batteries. *Electrochim. Acta* **2018**, *289*, 376–388. [CrossRef]

MDPI
St. Alban-Anlage 66
4052 Basel
Switzerland
Tel. +41 61 683 77 34
Fax +41 61 302 89 18
www.mdpi.com

Applied Sciences Editorial Office
E-mail: applsci@mdpi.com
www.mdpi.com/journal/applsci

9 7 8 3 0 3 9 4 3 3 5 0 6